GLOBAL CLIMATE CHANGE: NATIONAL SECURITY IMPLICATIONS

Carolyn Pumphrey

Editor

May 2008

*****

*****

Comments pertaining to this report are invited and should be forwarded to: Director, Strategic Studies Institute, U.S. Army War College, 122 Forbes Ave, Carlisle, PA 17013-5244.

*****

All Strategic Studies Institute (SSI) publications are available on the SSI homepage for electronic dissemination. Hard copies of this report also may be ordered from our homepage. SSI's homepage address is: *www.StrategicStudiesInstitute.army.mil.*

*****

The Strategic Studies Institute publishes a monthly e-mail newsletter to update the national security community on the research of our analysts, recent and forthcoming publications, and upcoming conferences sponsored by the Institute. Each newsletter also provides a strategic commentary by one of our research analysts. If you are interested in receiving this newsletter, please subscribe on our homepage at *www.StrategicStudiesInstitute.army.mil/newsletter/.*

ISBN 1-58487-352-3

CONTENTS

PREFACE

On March 30-31, 2007, the Strategic Studies Institute (SSI) and the Triangle Institute for Security Studies (TISS) held a colloquium on "Global Climate Change: National Security Implications." The 2-day event took place in Chapel Hill, North Carolina, and was well-attended by both academics and members of the U.S. Government and the Armed Forces.

This edited volume is based on this event. It reflects, as closely as possible, the form and content of the conference. Each chapter is based on a conference panel. The final chapter contains a keynote talk by General Paul Kern (USA-Ret.) and concluding remarks by Dr. Richard Weitz. The first seven chapters each contain three essays, a comment, and a discussion. The essays are not exact reproductions of the talks given at the conference. They remain true in spirit and length to the originals but have been adapted to be read. The question and answer sessions have been abbreviated in the interests of space. The reader who would like to hear exactly what was said at the conference is invited to turn to the video-taped proceedings which are posted on the TISS web site at *www.tiss-nc.org*.

The linkage between climate change and national security has received increasing attention over the last year, and the colloquium provided some timely insights. TISS would thank all those who made it possible. Too many individuals and institutions provided conceptual and other help along the way to be enumerated here. But we would like to acknowledge a few in particular. First, we owe thanks to the U.S. Army War College (USAWC) both for generous financial support and for recognizing this as a subject worthy of serious intellectual discussion. Second, we owe thanks to the participants. Their collegiality and professionalism made organizing this event a rare pleasure and the conference itself a success.

Besides these, the editor would like to express personal gratitude to a few individuals: Dr. Timothy McKeown, UNC-Chapel Hill, who first suggested that TISS look at the security implications of climate change; Dr. Kent Butts, USAWC, who insisted in the face of the skeptics that such a conference would be worth doing; Dr. Alex Roland, TISS Acting Director, who was instrumental in framing the agenda; and Dr. Douglas Johnson, SSI, who assisted from start to finish. Thanks also go to Michelle Koeneke for her invaluable assistance in organizing the conference and to Joseph W. Caddell, Jr., who read this manuscript. Last but not least, thanks go to Lowell Pumphrey, who wrote his senior thesis on the "Economics of Power Alcohol" (Princeton, 1936). He lived to learn all about the conference, though not to see the completion of this book. This volume is dedicated to him.

CAROLYN PUMPHREY, Ph.D.
Program and Outreach Coordinator
Triangle Institute for Security Studies

INTRODUCTION

Carolyn Pumphrey
Triangle Institute for Security Studies

The Evolution of a Problem.

Until fairly recent times no one thought climate changed, let alone was influenced by human activities. By the 19th century, scientists were theorizing that temperatures were affected by what we now call greenhouse gasses. And in the late 19th century, the Swedish scientist Arrhenius suggested that human industry might cause the planet to warm. But this notion was generally scoffed at. Over the course of the 20th century, the scientific community gradually came to terms with this theory and began to regard climate change—even rapid climate change—as more than a distant possibility.

Interest in climate change as a national security issue developed even later. Although the Central Intelligence Agency (CIA) did commission a study to look into the security implications of climate change in the late 1970s, the issue had little resonance until the late 1990s when the Senate Armed Services Committee declared that environmental destruction, including global warming, was "a growing national security threat." The Intergovernmental Panel on Climate Change (IPCC) was created in 1995 in part to allay fears. And then, in 2003, the rather notorious report commissioned by the Pentagon, "An Abrupt Climate Change Scenario and its Implications for United States National Security," provided a worst-case scenario, which suggested that climate change might have a catastrophic impact, leading to rioting and nuclear war.[1]

The State of the Problem Today.

So where are we in our thinking today when it comes to the science of climate change? There are still dissenting voices, and we cannot speak with absolute certainty. But science, we should remember, is essentially a culture of doubt.[2] As Karl Popper wrote at the start of the 20th century, "I think that we shall have to get accustomed to the idea that we must not look upon science as a 'body of knowledge', but rather as a system of hypotheses, or as a system of guesses or anticipations that in principle cannot be justified, but with which we work as long as they stand up to tests, and of which we are never justified in saying that we know they are 'true'. . . ."[3]

Nonetheless, the idea that there is such a thing as climate change is as close to established scientific fact as one can get. At its last meeting in February 2007, the IPCC concluded that human activity has indeed increased global atmospheric concentrations of carbon dioxide, methane, and nitrous oxide. It further concluded that "warming of the climate system is **unequivocal,**" and "most of the observed increase in globally averaged temperatures since the mid-20th century is **very likely** due to the observed increase in anthropogenic greenhouse gas concentrations." It is important to remember that the IPCC is an inherently conservative body. It can only make a statement by the unanimous consent of all the scientific representatives of all the world's governments. And it uses its words very precisely—so when it says "unequivocal," we know that it means exactly 90 percent certain—which is very certain indeed.[4]

What complicates the problem when it comes to the science of climate change, however, is that the devil is in the details. Although our instruments are improving and our data base enlarging, scientists have come up with many different scenarios as to how changes will play out over the next century. Timing is a case in point. Most estimates suggest a somewhat gradual timeline for change. However, there are some who fear that our current estimates fail to take into consideration what may happen if crucial tipping points are reached. If, for example, the Siberian tundra melts and releases its methane, this could act as a catalyst to climate change and make things happen a lot faster than expected. Some scenarios envisage sea-level rise sufficiently great to end civilization as we know it.[5] While we may acknowledge that these outcomes are less likely than some others, we ignore such possibilities at our peril.

The Consequences of Climate Change.

A level of uncertainty also exists when it comes to the correlations we can make between climate change and human security. The initial connections are easy enough to establish. For example, it is well known that warming facilitates the propagation of certain harmful bacteria and the spread of disease.[6] It is also clear that higher temperatures will lead to droughts which will affect agricultural production, and that ice melt will cause flooding especially in coastal areas. But it is much less easy to predict how these changes will affect different societies. Suppose climate change brings droughts and floods. Societies will cope more or less well depending on a lot of other variables. How adaptable are they, how effective are their political organizations, and do their cultural traditions serve as an obstacle or an aid to enlightened change?[7]

If you go one step beyond this and try to draw a correlation between these consequences of climate change and violence and conflict, you encounter a still more controversial area. Social scientists are not in agreement here. To be sure many—and possibly the preponderance—of social scientists think that such things as poverty and resource shortages lead to conflict, make post-conflict reconciliation harder, and provide a breeding ground for, if not terrorists, at least their supporters. However, there are informed and thoughtful minorities who disagree. Similarly, the historical record makes it quite clear that when life becomes unsustainable, people will simply get up and leave their homes, sometimes in large numbers.[8] But if mass migrations are a likely outcome of climate change, the precise ways in which they may lead to conflict are not clearly understood. Much will depend on the age of the migrants, the environment into which they move, and a host of other variables.

So, in short, we can assert with a large degree of confidence that the climate is changing, and that this has the potential to do us harm. But our challenge is how to approach this problem in light of the very real uncertainties.

National Security and the Climate Change Threat.

The purpose of the Triangle Institute for Security Studies (TISS) conference was to consider the national security implications of climate change. Here we run into some definitional problems. The meaning of the term "national security" is not agreed upon.[9] A traditional view is that it is concerned with the preservation of state sovereignty (most especially its monopoly of force) and the protection of national interests. In recent times, however, some scholars have argued that national security should encompass more than this. They argue that human welfare is in and of itself a security issue.[10] Thus construed, such things as poverty, disease, and environmental degradation are security threats not just because they lead to conflict, but because they are in and of themselves violations of "human security." Other scholars have argued that to define security in this way is to strip the term of all real meaning. This clearly has implications for the discussion at hand. If one takes the broadest definition of security, the mere fact that we are polluting the environment is a national security threat. If one takes the narrow view, the national security implications are less immediately obvious.

Protecting our "national interests" also means different things to different people. According to the *DOD Dictionary of Military Terms*, "National security interests include preserving U.S. political identity, framework, and institutions; fostering economic well-being; and bolstering international order supporting the vital interests of the United States and its allies." But opinion is divided as to whether this goes too far or not far enough. When it comes to climate change, for example, it seems as if those who are likely to suffer most hardship are people living outside the United States and not necessarily in regions of great strategic concern to us. Some see humanitarian intervention as an important component of our national interest—a fulfillment of a moral obligation that validates what we stand for as a society; a way to generate goodwill, and a way to preserve stability. But many others disagree.

These different underlying assumptions about the meaning of national security affect our response to climate change. They shape both threat assessments and policy recommendations.

Evaluating the Threat.

That climate change poses some kind of national security threat—impending danger or harm—seems clear. However, the United States faces many threats. The very act of preparing to meet one kind of threat may mean that we will be less able to meet another. So we must prioritize. In the pecking order of threats, where does climate change stand?

Evaluations vary markedly. A quick look at the official positions adopted by the leaders of different countries will provide an indication of this. As far as Former British Prime Minister Tony Blair is concerned climate change is **the** biggest threat facing the world today.[11] By contrast, as far as the Czech Prime Minister Vaclav Klaus is concerned, a still bigger threat comes from the very environmentalists who try to deal with the problem. In their efforts to halt global warming, they fatally endanger freedom and prosperity.[12] The U.S. administration under George W. Bush has certainly not placed climate change high on the list of priorities. As to conference participants, they agreed that the threat was a dangerous one, compounded by a context of rapid population growth, increasing economic appetite, pockets of extreme violence, and global interdependence.[13]

Why this enormous discrepancy? In part, the answer is because climate change is going to affect different nations to different degrees and in different ways. Unfashionable though these terms may be, there will be "winners" as well as "losers." The Russians, for example, are likely to benefit from the melting of Polar ice and the opening of new maritime routes. And, on a more frivolous note, the English wine industry may challenge the French.[14] By contrast, according to the Climate Change Index, people living in such places as Djibouti, Egypt, Pakistan, and Cuba are likely to suffer from serious physical problems leading in turn to political destabilization.[15]

In part the answer is because national security organizations have so many pressing concerns to consider. While chronic instability of the sort that may come out of climate change is obviously worrying, clear and present dangers like the war in Iraq and Afghanistan and the looming threat from Iran are almost inevitably going to command more attention.

Where climate change ranks in the scheme of things also stems from differences in interpretation of the scientific evidence. Analysts are currently working out a range of estimates. Scenarios vary from the totally catastrophic to the mildly disruptive but ultimately survivable.[16] It is all too easy, given this uncertainty, to choose the interpretation which best fits one's own cultural predilections or fills one's rice bowl.

Climate Change as a *U.S.* National Security Concern.

What about the United States? Is climate change currently expected to bring catastrophic changes in weather? Violent social upheavals? Intractable strategic problems? Or should we expect more subtle changes, more manageable problems?

Here, too, scenarios vary. Some models suggest that the North American continent will be among those most significantly (and negatively) affected by climate change.[17] Others suggest considerably less dramatic impacts. They say we might expect some serious flooding of coastal areas and rather serious drought in the Southwest. We might also expect more extreme weather patterns.[18] In principle, there seems to be agreement that we have the means to cope with most of these eventualities. Our recent experience with Hurricane Katrina, however, demonstrates that we have not yet learned how to take advantage of our existing assets.[19]

Even if we dismiss the worst case scenarios and assume that we will be spared the worst of what climate change can bring, we should note that climate change does **indirectly** pose very real national security concerns. Take terrorism, for example. The "war" against terrorists is very high up on the current list of national priorities. And there is persuasive evidence that extremism draws strength from the presence of poverty and inequality.[20] While images of streams of displaced persons swarming across the border are likely exaggerated, we know less than we should about how to integrate migrants into our society.[21] In some parts of the world, significant population movements could further destabilize volatile regions which we have a profound interest in keeping peaceful. The Middle East, for example, is vulnerable to water shortages, and climate change promises to exacerbate this problem.[22] The United States will also certainly have to deal with a rapidly changing strategic picture which may challenge its efforts to preserve world-wide stability.

In short, climate change is likely to be a stress-multiplier, to exacerbate tensions, and to complicate American foreign policy in a wide variety of ways.

A Sluggish Response.

Climate change is, then, at the very least a national security **challenge** for the United States. How effectively have we dealt with it in the past? The short answer is, not well.

If the polls can be trusted, the American public is gradually beginning to believe that climate change is not simply the figment of imagination of overexcited environmentalists. A 2006 Pew study found that about 41 percent of Americans think that global warming is a very serious threat. However, they rank it well below other issues as a national priority, and they are not willing to dip into their pockets to find a remedy.[23] While some corporations are trying to find ways to be more energy efficient, many others are dragging their feet.[24]

The American military is more environmentally conscious than is widely recognized. To some extent, this is policy driven. The Energy Policy Act, for example, makes energy conservation on bases a requirement.[25] U.S. Army *Field Manual 3-100.4* stresses environmental stewardship.[26] However, interest in environmental issues goes beyond grudging acquiescence to orders given by civilian "bosses." Some regional commanders have insisted on environmental engagement in the face of resistance from above.[27] Environmental security and disaster prevention, response, and recovery are now looked upon as acceptable military missions in that they are viewed as essential elements of regional stability.[28] And agencies like the Army Environmental Protection Agency work hard to promote advances in this area. At the same time, the Armed Forces continue to be committed first and foremost to the warfighting mission.

What is signally lacking is planning at a national level and clear directives from above. Environmental security is not part of any existing National Security Act. DOD Directive 3000.05 may tell the Armed Services that stability and support operations will receive the same priority as combat operations, but does not allocate specific funds to give these mandates "teeth." Corporations are not given the kind of incentives which might drive them to change their patterns of behavior. In short, while many different organizations and individuals are taking constructive steps to deal with the climate change problem, these are not, as yet, part of a coordinated strategic plan.

What is also lacking is a commitment to international diplomacy. The United States refused to ratify the Kyoto Treaty and remains reluctant to agree to any international legislation that would significantly curtail its actions.[29] It could be a leader in this area, and that it most certainly is not.

Solutions.

As a threat comes closer, our options for dealing with it become more and more limited. In the case of climate, studies suggest that we have a narrow window of opportunity to make some meaningful changes before irreversible damage is done.[30] It is, therefore, imperative that we remove our heads from the proverbial sand. So why has this problem failed to get much traction despite warnings from the scientific community and from some members of the defense establishment? There are arguably three overriding explanations.

1. In the first place, the **uncertainty** of science undermines the political will to act. Politicians do not like to operate in an uncertain environment and are likely to put such issues on the back burner. And warning about climate change is difficult. Because the public has a somewhat hazy understanding about how climate change manifests itself and because specific predictions are bound at times to be wrong, it is all too easy to throw the baby out with the bath water.

2. Closely related to this is the **communications dilemma.** A lot of climate change is about creeping dangers.[31] These appear remote and, sadly, rather boring to a public that likes drama. But if we attempt to wake the public up—as Former Vice President Al Gore did in *An Inconvenient Truth*—we may distort or manipulate the evidence. This too can give ammunition to the skeptics.

3. Finally, the problem has failed to really get our attention because it is all too easy to see it as the **other man's problem**. If the correlation between industrialization and accelerated climate change is indeed true, there would be both logic and justice in our doing proportionally more than others at this time to remedy the problem.[32] The fact is, however, that in developed countries we have some confidence that our powers to adapt will outpace the problems created by climate change. We do not, therefore, see it as a problem that will affect **our** vital interests. And so our enthusiasm to deal with it is correspondingly less.

Getting Traction.

Intrinsic to the whole process of addressing climate change will be finding ways to overcome these entrenched attitudes. The uncertainty of science does not, in fact, have to hamstring us. Risk management tools are available to help us deal with decisionmaking in uncertain environments.[33] Simple models exist which can help us explain this to the public. A wise person insures his home both against highly probable if not very serious potential dangers, as well as unlikely but potentially catastrophic disasters.[34]

The public needs to understand, moreover, that climate change will not just affect the polar bear. It will damage the health of our children.[35] People also need to better understand the implications of globalization. Not all currently appreciate how our security is affected by what is happening elsewhere in the world.

At the same time, we need to make clear the positive benefits that will result if we deal with the problem. Historically, reforms have tended to be more long-lasting and effective when they have served not just an ideal purpose but quite functional ends. Thankfully, one of the characteristics of the climate change issue is that it does not just threaten us. It also offers us opportunities.

Opportunities.

Consider just a few of the concrete benefits that might accrue to us if we take steps to deal with climate change. First, to slow down the pace of global warming, we will have to reduce carbon emissions. This is not just a tiresome necessity. If we do find alternative sources of energy, we will be less dependent on foreign oil and increase our foreign policy options. If our military becomes more energy efficient, its logistical capabilities will be enhanced.[36] If we reduce pollution, our national health costs will be cut. If our businesses discover creative new technologies, they will prosper and jobs will be created.

Second, climate change is a global problem and one that can only be solved by cooperation. As such, it offers us a chance to foster partnerships and build trust. Shared environmental concerns can bring people around the negotiating table. Providing clean water to local populations can make troubled zones more secure for our troops. And assisting people afflicted by climate-related disasters can help restore our somewhat tarnished image abroad.

Third, we can use climate change to promote the kind of developments that will benefit us all. It would be to **our** advantage if more foreign militaries were attuned to the idea of serving their people. It would be to **our** advantage if, instead of failed or failing states, we were to find legitimate governments capable of meeting the needs of their citizens and able to respond to a crisis.[37] Behind the scenes assistance to governments struggling to establish credibility and military to military engagement on environmental issues can do much to bring about such changes.[38]

Strategies.

Climate change, as a security problem, needs to be addressed at multiple levels. First, there is the root problem—the changing climate. Second, there is the human misery it will engender—we are talking of such things as poverty, disease, displacement, and social inequality. Third, there is the instability and/or changing strategic picture that will spring from all of the above. Simply put, our response needs to encompass at least three things: slowing down the rate of climate change and preparing to adapt to changes that cannot be avoided; taking steps to alleviate social distress; and preparing to cope with potential conflicts.

To slow down climate change, we clearly need to engage in mitigation and adaptation efforts. The terms are somewhat loosely applied and can mean different things to different people. For our purposes here, let us say that **adaptation** involves finding ways to accommodate ourselves to what is going to happen. **Mitigation** is an attempt to lessen harm.

If we want to change the direction of the curve, as it were, and slow down global warming, we have a range of options. These run the whole gamut from sponsoring research into alternative forms of energy to funding birth control programs. It will call for public education, legislation, and constructive participation in international diplomacy. If we want to adapt to changes that we think will inevitably occur, we should think, among other things, in terms of developing effective new technologies—desalination plants, perhaps, if we face drought; or hydropoles, if we face floods. We need to improve our forecasts and warning systems.[39] We need to identify areas of high vulnerability and work with the people there to help them build institutions at a **local** level capable of meeting future challenges.[40]

To deal with those human security problems that threaten to provide a fertile soil for extremism and violence, it would seem logical to address the problem of growing social inequality. This might mean providing assistance to vulnerable nations. The United States can play a role here, as it has done in the past. It can fashion suitable international institutions to help create economic prosperity.[41] And it can provide states with the tools to adapt to increasingly stressful conditions. Fostering legitimate governments is a key element of this strategy.

To deal with future conflicts, more comprehensive planning is essential. The entire range of plausible threats needs to be delineated. Alternative approaches and cost-benefit analyses must be run to establish what can be done, when, and at what cost.[42] We need to improve our warning systems so that a warning is not simply given, but also heard and acted upon.[43]

Coordination at many levels is also going to be crucial. It has been identified for some time as one of our national shortcomings. Dealing with climate change generated conflict calls for contributions from a wide variety of agencies, local, national, and international; military and civilian. An effort needs to be made to determine what kinds of organizations are best suited to what kinds of activities so as not to duplicate efforts. Thought also needs to be given as to how to take best advantage of assets already in place: How, for example, can uneasy bedfellows like nongovernmental organizations (NGOs) and the military work together without arousing the suspicion of host countries? Along similar lines, effective interagency action may require new legislation and better definition of Department of Homeland Security authority.[44]

Finally, a precondition for success in all three cases must be improved understanding of the problem—better intelligence, better science, and better understanding of the relationships between such things as violence, society, and climate change. And plans, at every level, must get the ear of the leadership, so that they do not sit in elegant volumes and gather dust.[45]

Responsibility.

If climate change is ever to be successfully dealt with, someone is going to have to assume responsibility. Who should it be? Is responsibility related to culpability? Or is it related to capability? What organizations are best equipped to deal with particular tasks? What role, in particular, should U.S. Armed Forces play? As major consumers of energy, and as organizations that are periodically engaged in warfare, armed forces inevitably contribute to climate change in general and to global warming in particular. They also have assets and capabilities that many other elements within our society do not. Should they, therefore, take the lead when it comes to finding solutions?

The answer is both "yes" and "no." The Armed Forces—perhaps the Air Force more than the Army—can clearly help reduce global warming. Military research and development can be applied to the task of developing new and more efficient technologies. Military demand for fuel efficient vehicles could help stimulate similar demand in the private sector. As a huge organization, merely applying efficient technologies on bases[46] can (and in fact is) making a big difference. And, as noted above, developing fuel efficient systems has a military **utility**.

The Armed Forces may also be called upon to restore stability to regions devastated by extreme weather events or to provide humanitarian assistance. Peacekeeping and stabilization operations are in fact now officially part of the core mission of the Armed Forces.[47] And our military does have some unique advantages. It has a widespread presence throughout the world; is good at dealing with emergencies; and has vast, though not infinite, economic resources at its disposal. Its ability to gather and analyze intelligence, and the cultural awareness and linguistic skills of some of its special forces are among other less widely recognized assets that could be useful in disaster relief.[48] However, for a variety of reasons—traditional understanding of proper civil-military relations among them - it is probably best that it play a supporting rather than a leading role in such missions, plugging gaps where appropriate, but not remaining in charge.[49]

Furthermore, as noted earlier, the military must maintain the ability to deal with a wide range of contingencies. It has to make choices when it comes to structure and training. Gearing up for climate change may be compatible enough with efforts to deal with asymmetrical threats. But some fear that it may reduce its ability to get ready for war of a more traditional kind. This particular dilemma is far from resolved. However, even if we grant that the military should not be the leader in the struggle against climate change, it is still important to recognize that it is a vital contributor.

In the final analysis, however, it is clear that this is not just a job for the military. We also need a government that incorporates climate change science into its strategic planning and that leads international efforts to create partnerships and institutions capable of responding to the threat. We need businessmen willing to invest money and effort in clean, energy-efficient technologies. And we need ordinary citizens who are sufficiently well-educated on this subject to put the problem in perspective, pay carbon taxes if needs be, and buy the right sort of light bulbs. In short, it is a job for everyone.

Conclusion.

To conclude, climate change is certainly an issue that deserves serious consideration as a national security concern. It may not appear at the moment to be the most crucial threat facing the United States. However, we ignore this threat at our peril. We stand to lose a great deal if we do not move fast; the evidence suggests that the problems will increase incrementally if we wait. And at the same time, we stand to gain much if we do act—a healthier, cleaner environment, a more stable world community, better relationships with other countries, and greater national security.

ENDNOTES - Pumphrey

NOTICE: Endnotes referencing chapters are referring to this volume.

1. Spencer Weart, Chapter 1.

2. Richard Feynman, Nobel prize-winning physicist. "Religion is a culture of faith; science is a culture of doubt."

3. Karl Popper (1902-94), The Logic of Scientific Discovery.

4. Robert Corell, Chapter 1.

5. This worst case scenario is outlined in the study by the Center for Strategic and International Studies article by Jim Woolsey, cited by Richard Weitz, Chapter 8.

6. Andrew Price-Smith, Chapter 2.

7. Erika Weinthal, Chapter 2; Robert Dorff, Chapter 5.

8. Comment made by Brian Fagan, keynote speaker at the TISS-AWC Conference. Fagan, author of *The Great Warming: Climate Change and the Rise and Fall of Civilizations,* London: Bloomsbury Press, 2008, discussed the medieval warming period, showing the sometimes cataclysmic impact climate change has had on human societies.

9. National Security: "A collective term encompassing both national defense and foreign relations of the United States. Specifically, the condition provided by: a. a military or defense advantage over any foreign nation or group of nations; b. a favorable foreign relations position; or c. a defense posture capable of successfully resisting hostile or destructive action from within or without, overt or covert." DOD Dictionary of Military Terms.

10. Richard Matthew, Chapter 1. See also, for example, Stephen E. Sachs, "The Changing Definition of Security," available from *www.stevesachs.com/papers/paper_security.htm,* accessed November 2007.

11. At the G8 conference in 2005, British Prime Minister Tony Blair referred to Climate Change as "Probably long term, the single most important issue we face as a global community." Cited in *sciencepolicy.colorado.edu/prometheus/archives/climate_change/000047tony_blair_comments_.html*, accessed November 24, 2007.

12. See, for example, *www.ft.com/cms/s/2/9deb730a-19ca-11dc-99c5-000b5df10621.html*, accessed November 24, 2007. See Marvin Soroos, Chapter 4.

13. Matthew, Chapter 1; Timothy McKeown, Chapter 2.

14. CBS news article, September 25, 2006, available from *www.cbsnews.com/stories/2006/09/25/eveningnews/main2037991.shtml*, accessed November 24, 2007.

15. This is based on the Climate Change Index (CCI) and consists of three components each carrying equal weight: (1) Coastal explosure, (2) Inland exposure, and (3) Health exposure.

16. Henry Gaffney, Chapter 5; Weitz, Chapter 8.

17. According to the CCI, the United States is one of 31 countries (out of 189 profiled) that is extremely vulnerable to climate change. It ranks as the 15th most exposed to climate change impact (physical and health) with a CCI of 1.85 (with 10 being countries with least exposure, and countries at extreme risk falling between 0.0-2.5). It is also only one of three developed nations in this category. The Maplecroft Map is available on line at *www.maplecroftmaps.net*. Go to the "climate change" map to find the "Global Map of Vulnerability." The global map of climate uses data from the following sources: *Population, Landscape and Climate Estimates (PLACE) from the Centre for International Earth Science Information Network (CIESIN)*, Columbia University; the World Health Organization (WHO) publication "Comparative Quantification of Health Risks"; the *EM-DAT: The International Disaster Database.*

18. Joshua Busby, Chapter 3; Corell, Chapter 1.

19. Michael Glantz, Chapter 5.

20. Kent Butts, Chapter 3.

21. McKeown, Chapter 2.

22. Weinthal, Chapter 2.

23. Anthony Leiserowitz, Chapter 4.

24. Simon Rich, Chapter 4.

25. David Sheets, Chapter 6.

26. Karen Hulme, Chapter 6.

27. Dorff, Chapter 3, Q and A.

28. Bert Tussing, Chapter 7.

29. Dennis Tänzler, Chapter 4.

30. *Ibid.*

31. Glantz, Chapter 5.

32. This is, in fact, noted in the 1992 UN Framework Convention. See Tänzler, Chapter 4.

33. Weart, Chapter 1.

34. Corell, Chapter 1.

35. Leiserowitz, Chapter 4.

36. E. Thomas Morehouse, Chapter 7.

37. Dorff, Chapter 5.

38. Tussing, Chapter 7.

39. Glantz, Chapter 5.

40. Dorff, Chapter 5.

41. McKeown, Chapter 2.

42. Douglas V. Johnson II, "Global Climate Change: National Security Implications," Colloquium Brief, U.S. Army War College and Triangle Institute for Security Studies, 2007.

43. Glantz, Chapter 5.

44. Johnson, "Global Climate Change."

45. Christopher Schroeder, Chapter 6.

46. David Sheets, Chapter 7.

47. DOD Directive 3000.05, "Military Support for Stability, Security, Transition, and Reconstruction (SSTR) Operations," November 28, 2005.

48. Tussing, Chapter 7.

49. *Ibid*.

CHAPTER 1

DEFINING THE PROBLEM

The purpose of this chapter is to put the issue of global climate change into perspective, define terms, and introduce the reader to major debates and areas of uncertainty. Dr. Spencer Weart shows how global warming has come, over time, to be framed as a national security issue. Dr. Robert Corell lays out the scientific evidence for climate change in general and global warming in particular. Dr. Richard Matthew evaluates global climate change as a security threat and outlines some of the definitional problems that face us.

A National Security Issue? How People Tried to Frame Global Warming

Spencer R. Weart
American Institute of Physics

Global Warming as a Scientific Puzzle.

Nice weather we're having, isn't it? People are, and always have been, interested in the weather and changes in the weather. Climate was something that came with the territory—weather was what changed. People were very concerned of course, about floods, droughts, cold spells, and so on. These were things which came and went, and went and came again. Climate, by contrast, wasn't supposed to change. It came as a great surprise in the 19th century when people found that there was a historic record of climate change, namely the ice ages. The ice ages were a great puzzle and one of the big intellectual challenges of the 19th century. One of the leading lights of the Swedish Academy of Sciences, Svante Arrhenius, thought that he had an explanation for the ice ages: They were mainly caused by carbon dioxide, what we now call a "greenhouse gas."

Over geological ages, carbon dioxide was put into the atmosphere by volcanoes. It was taken out of the atmosphere by chemical absorption in rocks or, especially, by dissolving into the oceans. If something happened to change this, if there were fewer volcanoes, for example, there would be less carbon dioxide in the atmosphere. Arrhenius made some calculations and concluded that if you cut the amount of carbon dioxide in the atmosphere in half, that would lower the temperature enough, probably, to bring on an ice age. This was not an entirely new idea. His calculations were new, but the idea had been around since the mid-19th century. John Tyndall, an English scientist, explained it neatly—you'll never get a better explanation than this of the greenhouse effect: "As a dam built across the river causes a local deepening of the stream, so our atmosphere, thrown as a barrier across the terrestrial rays [that is, the heat rays, the infrared radiation coming up from the surface] produces a local heightening of the temperature of the Earth's surface."[1] Adding carbon dioxide high in the atmosphere, will block more of the radiation coming up from below, causing an imbalance—more radiation coming in than going out—so that the temperature underneath has to rise until there is enough radiation coming out to restore a balance.

A colleague of Arrhenius pointed out that humanity was putting as much carbon dioxide into the atmosphere as volcanoes, more or less. Over the 19th century, industry had grown to a prodigious extent, with puffing smokestacks dominating the cities—which, by the way, was regarded as a welcomed and happy sight, standing for employment, industry, progress, overall a very good thing.

So Arrhenius decided to calculate things the other way: What happens if we double the carbon dioxide level of the atmosphere? A very rough calculation, which only took him a year with pencil and paper, suggested that it could raise the temperature by several degrees. In chilly Sweden, that sounded like a good idea. "We may hope to enjoy ages with more equable and better climates," he wrote, "ages when the Earth will bring forth much more abundant crops than at present. . . ."[2] He did not expect this to happen for a very long time, thousands of years in the future. So at the outset, global climate change was not framed as a problem for our society; it was framed as an interesting scientific problem. And even as a scientific problem, the main interest was not in global warming but global cooling. Some thought the greenhouse effect might provide the long-sought solution to the puzzle of what caused ice ages.

When Arrhenius presented the scientific world with his speculation that human industry might some day warm the planet, the idea was scoffed at. How could humans produce an effect great enough to affect the climate? And of course if they did, would it not be all for the better? At the time, it was generally held that we lived in a balance with nature, and the balance of nature would make sure nothing terribly bad happened. You could even take it to the extent of believing in a kind of homeostasis—there was a natural system that regulated everything, kept everything stable. The oceans after all, would absorb carbon dioxide as fast as the gas was produced, and everything would be stable and fine.

A few began to raise doubts in the 1930s, for climate changes were beginning to be visible. People had accumulated enough good weather records by that time to see that there were some long-term changes underway in the climate in Africa and the United States and so on. For the northern hemisphere, the record of good temperature measurements began in the 1880s. This was far enough back for people to see that there had been a slow increase. As one of the news magazines put it, "Gaffers who claim that winters were harder when they were boys are quite right . . . weather men have no doubt that the world at least for the time being is growing warmer." Still, if some people recognized that climate did change, they saw that as just a part of some natural cycle: "Meteorologists do not know whether the present warm trend is likely to last 20 years or 20,000 years." Such natural cycles seemed to have nothing to do with human emissions. There was one man who challenged this, a well-known engineer but an entirely amateur meteorologist, Guy Stewart Callendar. Looking over historical data, he concluded that the carbon dioxide level in the atmosphere had in fact been increasing. He predicted that the increase in temperature due to the artificial production of carbon dioxide would come to 0.3 degrees Celsius per century. Again, Callendar saw nothing very urgent in this. He saw it largely as a scientific problem in geophysics, and mainly of interest if it could explain ice ages. To other scientists, it was only one theory of climate change, and not as likely as many other theories going around, having to do with sunspots and other solar changes, or variations in ocean currents, or who knows what. As a science writer put it, "Everyone has his own theory—and each sounds good—until the next lad comes along with his theory and knocks the others into smithereens." The carbon dioxide greenhouse effect theory of climate change was just one more thing on the shelf with all the other bric-a-brac and not by any means the most attractive.

The idea that humans could change the climate began to change with the coming of atomic energy. The atomic bombs themselves showed that humanity was now in fact in possession of forces of geophysical magnitude. Also, as fallout went around the world, people began to realize that what you do in one place could have an effect on the environment far away. Some people even began to say that bomb tests were affecting their weather, causing more floods or droughts. If weather records said the world was getting warmer, a news magazine reported that "large numbers of people wonder whether the atomic bomb is responsible for it all." If theoretical ignorance remained complete, there was an increasing willingness to accept that human civilization might have some impact on the natural world. Global warming began to be reframed, as something that might have more than an abstract scientific interest.

First Worries about Climate Change.

Now we begin to get into some of the real science. For lack of time, I will just tell a couple of stories that relate to how the matter was framed. One of the stories starts with Roger Revelle, an oceanographer who had been studying sea water since the 1930s. Sea water was of interest to the U.S. Navy. By the end of World War II, Roger was Commander Revelle, in charge of a large scientific team sent to the atoll at the 1946 Bikini atom bomb test, to study what effects the explosion had on the sea water and the lagoon. This revived his interest in sea water chemistry. Now, sea water is not just salt water; it is a complex chemical solution. In fact it is a buffered solution, and Revelle realized that means it does not take up carbon dioxide as rapidly as people had thought. Moreover, he was aware that the human population had doubled since Arrhenius's day, and the output of industry per capita had more than doubled. More important still, he was aware (what earlier people had failed to recognize) that this was an exponential trend, which was probably going to double and redouble again. So Revelle took the trouble to do a new calculation, taking into account his understanding of the lesser capacity of oceans to absorb carbon dioxide and the ever increasing emissions of CO^2. In 1956 he came to the conclusion that, in fact, the carbon dioxide content of the atmosphere must increase.

In 1956 Revelle testified to Congress that the Arctic Ocean might become navigable. If so, he said, "the Russians will become a great maritime nation."[3] This is the first statement that I have found that hinted at the national security implications of climate change. Revelle was actually talking through his hat, speculating about something that he didn't actually expect to happen for centuries. The reason he was testifying (before an appropriations committee) was to ask for money to fund new research, planned for the forthcoming International Geophysical Year (IGY 1957-58). In the 1950s, the word "Russia" was a common code word for "give us some money." Revelle's personal attitude was expressed better in a statement he made several times, that by adding carbon dioxide to the atmosphere "human beings are now carrying out a large scale geophysical experiment." In short, it was scientifically interesting, and probably his main interest was the old one of hoping to explain what caused the ice ages. He was little concerned with a future rise in temperature.

Because of the scientific interest and because he did get money for more research during the IGY, he hired a young post-doctoral candidate, Charles David Keeling, to study carbon dioxide. Keeling went to Little America, the base the Navy established in Antarctica for the IGY. In the very pure air there, he carried out the most meticulous measurements that had ever been made on carbon dioxide. Within 2 years, he was able to show that, exactly as Revelle had predicted, the carbon dioxide level was increasing.

Whether that meant that the temperature would increase, nobody was sure. But if it did, then as one scientist put it, "there would seem to be every reason for producing as much carbon dioxide as we can manage. It is helping us towards a warmer and drier world." Such optimistic views about technological "progress" had begun to shift, however, and not only because of atomic weapons—although that certainly had something to do with it. Environmentalism had come on the scene. (Earth Day 1970 is often cited as a breakthrough in public opinion and political clout.) And weather changes were worrying now. In the early 1970s there were very severe droughts in Africa, the American Midwest, and elsewhere. Concern about the world food supply grew with the collapse of the Russian wheat crop. In short, a variety of things stimulated doubts about the benevolent impact of climate change and led people to consider that human-generated pollution might possibly be affecting climate, at least regionally.

Concern about atmospheric pollution was no longer limited to complaints about smog in cities. Industrial haze was found to be spreading around the entire Northern Hemisphere. Environmentalists took note. In the long list developed in the 1970s of harms we were doing to the environment—deforestation, overgrazing, acid rain, smog—global warming took a modest place. Global warming became framed as an environmental issue. It was a fateful move, for environmentalism in general was increasingly associated with only one part of the political spectrum, the left.

Speculating about Catastrophe.

Meanwhile, temperatures in the Northern Hemisphere had begun to turn down. In the Southern Hemisphere, we now know, this wasn't happening, but they didn't have good records for the Southern Hemisphere at the time. One possible explanation for the cooling was the smoke and other particles that industry and so forth were putting into the atmosphere. One scientist, Reid Bryson, proposed that what he called the "Human Volcano" was putting so much soot and dust into the atmosphere that it was producing a cooling effect around the world. (Volcanoes pour out not only carbon dioxide gas, but also smoke and sulfate particles, and a great eruption will temporarily shade the planet's surface and cool it.) This bothered Bryson because there was a lot of data accumulating, including data he had found himself from studies of ancient tree rings and so forth, that climate change could be quite rapid and persistent. In the American Southwest and Midwest, he found signs that there had been very, very severe droughts that had come on in less than a hundred years—it was no longer a question of ice ages that took ten thousand years to settle in, or even a thousand years. It seemed like serious climate change could come within as little as one or two human generations. He wondered whether, by putting all this in the air, we might be bringing on a new ice age.

If that was so, there would be serious implications. As *Fortune* magazine put it in an article inspired by Bryson, we might see "a billion people starving" or other very serious consequences. (In fact, recent work suggests that pollution and other human influences did play a role in the great drought that hit Africa at just that time, the early 1970s, starving millions and killing tens of thousands.) Now, while this kind of talk was found in the popular press, if you look at the refereed scientific literature, you will find that nobody was saying anything so radical. Some scientists were saying that there might be a rather gradual cooling. Others were saying that it was more likely that greenhouse warming was coming on. The important new scientific idea was that serious climate change could be relatively rapid, within a single century, and it could be triggered by humans.

These ideas provoked a study commissioned by the Central Intelligence Agency on the potential implications. The study built on Bryson's ideas, suggesting that there might be serious problems—what we might well call national security problems—coming from climate change. Indeed they predicted that in the worst case, "there would be increasingly desperate attempts on the part of powerful but hungry nations to get grain any way they could. Massive migrations, sometimes backed by force, would become a live issue. . . ."[4]

Now, this was only one small report, by no means an official government position. Probably the dominant view among meteorologists was that, as one authority put it, "the climatic system is so robust . . . that man has still a long way to go before his influence becomes great enough to cause serious disruption."[5] The refereed scientific literature had it all down as a matter that needed, and deserved, further research. The old assumptions about the balance of nature remained widespread: Maybe we'll get warmer, maybe we'll get cooler, but more likely nothing will change very badly or very fast.

Scientists Begin to See a Real Problem.

To explain how this view changed, I have to skip back in time to another line of the scientific story. Let's visit the meteorology group in Princeton in 1952. The leader is Jule Charney, the first man to work out the mathematics for how to predict weather on a computer. His main colleague was Norman Phillips, the first person to actually program a computer that could predict weather 24 hours ahead. It took a 24-hour computer run, but it was a start. In fact the most important member of the team was not human: *MANIAC 1*. This computer was designed specifically to study the dynamics of compressible fluids, although the compressible fluid originally in mind was not the atmosphere, it was the plasma in a hydrogen bomb. I'm making a little point here about the importance of military funding in this entire story.

Jumping forward in time to the mid 1960s, the next main figure is Syukuro Manabe, working in another lab in Princeton. Manabe took the weather computer models, which were working pretty well by that time, and developed a model for the average weather over a year, that is, climate. Compared with the real planet, Manabe's world wasn't too bad. For the United States, for example, his model had a lot of precipitation in the Pacific Northwest, not much in the Southwest. He had the tropical rain bands, a wet Brazil, a dry Chilean desert, and so on. In short, his computer was simulating the real world to a rough approximation.

Now Manabe decided to put some extra greenhouse gas in his model atmosphere: Let's double the carbon dioxide and see what happens. The temperature went up several degrees. And in fact, this is a robust feature of models to this day. You cannot make any kind of mathematical or computer model of the climate, which will (1) produce a climate that looks anything like the real world's climate, and (2) not have it heat up if you add carbon dioxide. That's just the way models work.

This began to catch the attention of physicists, including a group of elite physicists who called themselves the Jasons. One summer they undertook to study the long-term impact of atmospheric carbon dioxide on climate. Some of you may be familiar with the Jasons, whose main activity was and remains wide-ranging studies for the military. After decades of working out of public view for the Defense Advanced Research Projects Agency (DARPA) and other defense agencies, they came under great fire from their fellow academics during the Vietnam War. When they decided to study climate, they did it for the Department of Energy. It was a step away from national security affairs; they wanted to do something more civilian in nature, so they took up the study of climate as a break from their usual military-oriented summer studies. But what they concluded was, perhaps, closer to national security than they might have thought.

After devising their own computer model, they got the same result Manabe and others were getting: They predicted a warming of a few degrees Celsius by the middle of the 21st century, and much more warming in the Polar regions. That Arctic warming, by the way, is a robust feature of climate models; even Arrhenius got that with his pencil and paper. The Jasons concluded that there might be serious consequences for the world's food supply. There would be a sea-level rise, which could possibly be damagingly rapid. As a result of the climate changes, they reported, there could be a large-scale displacing of populations. Of course populations have always been moving from one climate zone to another, as when Mexicans come to the United States, but they might not be happy to have it imposed upon them. The Jasons were also aware of, shall we say, the unknown unknowns, a variety of other possible effects: acidification of the oceans and so forth. In short, pretty much the whole range of potential problems that we are now concerned about were already well-understood by physicists by the early 1970s.

This was just a bunch of physicists doing a study out of their own curiosity. But their report did cause greenhouse warming to get attention, for the first time, at the highest levels of government. The President's science advisor was a geophysicist, Frank Press, and he decided to have a real study done by a panel of the National Academy of Sciences. It was headed by Jule Charney, who had become (and remains) the grand old man of computer weather prediction, but he had no fixed views on climate change. Charney and his panel studied the entire question carefully and concluded: Yes, if you doubled atmospheric carbon dioxide, there would be a 1.5 to 4.5 degree Celsius rise. This is not as precise as it sounds; what they were really saying was "roughly three degrees, plus or minus 50 percent." It was a good estimate even so, not far from current values (those computer models really are robust). But carbon dioxide would not double until well into the next century, and the 21st century seemed very far away at that time. So when it came to actual consequences, the Charney Panel only needed to conclude that "the socioeconomic consequences may well be significant, but . . . cannot yet be adequately projected." There was plenty of time for more research, and that was the only action they recommended.

The research did proceed, and it gradually shifted opinions in the scientific community. What had seemed a rather vague and remote possibility began to look more and more serious and certain. I will just mention one of the many lines of research: the Soviet Union's last great achievement. At Vostok in Antarctica, the most remote and barren spot on the planet, they drilled kilometers down through the ice. Analysis of gases trapped in the layers of ancient ice showed that carbon dioxide and temperature went up and down together through the last ice age (in fact, the core eventually got deep enough to record the same thing through the last four ice ages). In short, there was demonstrably a tight feedback between carbon dioxide and temperature. Raise the temperature and the carbon dioxide will go up, raise the carbon dioxide and that will raise the temperature still more, bringing out yet more carbon dioxide, and so forth—a strong and rather frightening feedback.

Getting Serious about Global Warming.

Meanwhile, the temperature in the Northern Hemisphere had begun to rise. A few scientists began to feel it was time to alert policymakers and the public that a real problem could be foreseen. The pioneer was Jim Hansen, who already in the early 1980s had predicted, quite accurately, that the "signal" of greenhouse warming would emerge from the noise of normal climate fluctuations sometime around the year 2000. Testifying before Congress in 1988, the hottest summer ever in Washington, DC, he told people, "It's time to stop waffling and say the evidence is pretty strong that the greenhouse effect is here."[6]

Congress took little notice (aside from Al Gore and one or two others). Environmentalists did, however, begin to join the scientists in calling for national policy attention to global warming. It was, in fact, the new Environmental Protection Agency that issued the first official government report on climate change. As the *New York Times* described their findings, "Global warming caused by industrial pollutants is likely to shrink forests, destroy most coastal wetlands, reduce water quality and quantity in many areas, and otherwise cause extensive environmental destruction in the United States over the next century."[7] Other environmentalists went farther. For example, one publication exclaimed that "A climate change that turned the Great Plains of North America into an arid zone would be analogous to a major military disaster."[8] Thus environmentalists framed climate change as a problem on the same order as national security issues. Their plan, typical of the 1970s, was to take money away from the military. They said we should spend less money on the military and use it to address the more serious environmental threats to our well-being. Other government agencies gave all that little credence. But the pressure became more serious with the end of the Cold War, when the military lost what had been its primary mission.

That prompted Senator Sam Nunn and other members of the Senate Armed Services Committee (Al Gore for one) to bring the environment into their deliberations. In 1990 they tried to stir up public interest with a joint declaration that environmental destruction, including global warming among other problems, was "a growing national security threat." Their conclusion was that we should give more money to the military, or at least stop cutting the military budget so the armed forces could do something about the environment—

and, not coincidentally, preserve the prerogatives and power of the Armed Services Committee.

This idea of going environmentalist was not very natural to Senator Nunn, and it struck no resonance whatsoever with then Secretary of Defense Dick Cheney. Moreover, there was a strong political counter movement against any worrisome talk about the environment. An attack specifically aiming to reduce concern and action on global warming was mounted by the fossil fuel industries and their allies. They sponsored scientific and quasi-scientific reports, and spent literally millions of dollars on press releases, websites, and lobbying the administration and Congress. They had considerable success convincing people that the science of climate change was nonsense, or at any rate so dubious that it was nothing anybody really needed to worry about, still less act upon. Policymakers, partly because of these countervailing forces and partly from the usual inertia about undertaking anything new, were perfectly happy to do nothing about climate change.

Meanwhile the science continued to progress. Again, I can only take one example from many hundreds of significant findings. In 1991, a volcano in the Philippines erupted and threw a cloud of sulfates the size of Iowa into the stratosphere. Jim Hansen saw this was a fine natural experiment: He could put the cloud of sulfates into his computer model and see what happened. The models were much better now, and he could follow climate month by month. His results predicted what the effects of the Pinatubo eruption would be, in what regions the sulfates would produce some cooling for a year or two until they washed out of the atmosphere. The world followed his simulation with impressive accuracy. This is one of many examples where computer models faced a severe test and passed it with flying colors.

Meanwhile, as Hansen and others had predicted, the temperature continued to rise—indeed, now clearly rising beyond the range of normal variability. Warming was greater in Arctic regions and had other particular features that matched the predictions from the greenhouse effect. No such neat match to the observations was found by people who tried other explanations for the warming, such as a change in the sun or just random variations. It was increasingly clear that people like Hansen who had predicted warming were right, whereas the skeptics' predictions had all failed. Policymakers finally had to take action.

Seeking Agreement on the Threat.

The outcome was the Intergovernmental Panel on Climate Change (IPCC). It was designed by the Reagan administration, primarily as a replacement for the self-appointed committees of scientists (like the National Academy of Sciences' panels), which the administration thought were unduly alarmist. The IPCC was designed so that it can make a statement only by the unanimous consent of all the scientific representatives of all the world's governments. This is surely a recipe for conservatism, if not paralysis. In fact, the first statement that the IPCC issued in 1995 was perhaps the most weasel-worded statement ever issued by a scientific body: "The *balance* of evidence *suggests* that there is a *discernible* human influence on global climate . . . [emphasis added]"[9] And yet it does say something, a serious something.

By now, the debate on global warming was well underway. But the most important debate was not the one most people heard about, what we might call Greenpeace versus the editors of *The Wall Street*

Journal. The real debate was conducted under the auspices of the IPCC, and it engaged all of the world's thousands of climate scientists in round after round of research, debate, workshops, working papers, panel discussions, and more research. Surprisingly in the end, the IPCC, after a fierce night-long debate among the representatives, was able to come out in 2001 with a truly meaningful consensus statement. And it was scary, "Temperature is very likely to increase by 1.4 to 5.8 degrees C by 2100 . . . a rate without precedent during at least the last 10,000 years." So that, essentially, was the close of the debate, at least among intellectually informed people who did not have a preconceived bias. (Of course, a very few senior scientists could not bring themselves to abandon views they had formed decades earlier. That propped up the fixed opinions of people with less grasp of science, but who felt that to admit that anything might require government regulation would be an offense against their ideology or their wallet.)

The consensus that serious warming was, if not certain, then surely very likely, was not the end of the research. Again I will just mention one of the many developments. Ice cores were gathered, at substantial risk to life and limb, in places from Greenland to the Andes. Where the snowfall is heavy enough, you get annual layers and you can study them like tree rings. In some places, the changes of past climate were shockingly rapid. Such evidence convinced the National Academy of Sciences to convene another panel to look into rapid climate change—more often now called abrupt climate change. They found that a 3 degree Celsius (that is, 5 degrees Fahrenheit) regional change is possible within 1 decade. That would be a great catastrophe for the region affected. This new scientific thinking, they said, "is little known and scarcely appreciated." I think that is still to some extent the case (the IPCC reports in particular give little attention to these less likely, but vastly more dangerous, scenarios).

However, the concern did to some extent reach the public. It also reached the attention of people in the Pentagon and some defense intellectuals, who commissioned a 2003 report on "An Abrupt Climate Change Scenario and its Implications for United States National Security."[10] The key word here is scenario—in true military think-tank fashion, they decided to study a worst-case scenario, just so we could be prepared. When it was leaked to the press, it inspired some sensational stories: "Climate change will destroy us. Secret report warns of rioting and nuclear war. Threat to the world is greater than terrorism." (We have heard a similar statement recently from some high authorities in Europe.) The most frightening possibility, they said, would be a collapse of the North Atlantic ocean current system, in which case "Britain will be Siberia in less than 20 years."[11] That was also the premise for a widely seen summer disaster movie, "The Day After Tomorrow." This was definitely a worst-case scenario; all scientists agree that there is virtually no chance of such an event within the foreseeable future . . . that is, the next century or two.

While all this got a lot of attention, there was also push-back. Every winter somebody publishes an editorial cartoon joking about a meeting on global warming postponed on account of snow or whatever. These always appear in conservative media, for already by the time of the Reagan years, the issue had become strongly politicized. In terms of the degree of worry about consequences of global warming, a Gallup poll just released shows that among Democrats the level of worry is about 75 percent; among Independents, it is 60 percent; and among Republicans, it is 35 percent. So if you tell me how you vote, I can make a good guess about what you think about a scientific issue like climate change (a sorry indictment of the intellectual state of our society right now).

Yet opinions were shifting even among conservative business leaders, who began to understand that climate change could have serious and direct economic consequences. The unprecedented European heat wave of 2003, which killed 10,000 people in France alone, certainly had a great impact on public opinion in Europe. It is now believed that this heat wave was made considerably more likely by global warming, and that more will increasingly follow. As for Hurricanes Katrina and Rita that struck the Gulf Coast in 2006, nobody knows whether they were made more likely by global warming. The best guess is that such disasters were about equally likely to happen in 2006, 1906, or 2106 (other changes in weather, like increased floods and droughts, are much more certain). However, the hurricane catastrophes did show Americans what is meant by "higher sea level" and "environmental refugees." The destruction of New Orleans was a striking illustration of the kind of forces that we expect will in fact drive millions of people around the world from their homes.

Coming back to where the big scientific picture now stands, let's recall the IPCC's 2001 statement that future warming of several degrees was "very likely." Just last month, after half a dozen years of intense research and discussion, they issued their next statement. And their 2007 statement says, in effect, "Like we said . . . !" The temperature will rise by more than one degree Celsius, perhaps as much as six (with a small but awful possibility of even more). And the changes predicted to come, the heat waves and droughts and floods, the rising sea level, and stress to countless species, are already being observed around the world. At the request of the British government, a team of experts under a former Chief Economist of the World Bank studied a businesslike "worst case worth insuring against" for climate change in this century (not as bad as a national security worst-case scenario). They concluded the impacts could easily be as serious as the consequences of World War II.[12]

So there you have, in 20 minutes, 20,000 person-years of intellectual endeavor.[13]

ENDNOTES - Weart

1. J. Tyndall, "Further Researches on the Absorption and Radiation of Heat by Gaseous Matter," in *Contributions to Molecular Physics in the Domain of Radiant Heat,* New York: Appleton, 1873, p. 117.

2. He thought this increased productivity would take place in part because increased carbon dioxide would fertilize the crops.

3. Testimony of Roger Revelle, U.S. Congress, House 84 H1526-5, Committee on Appropriations, *Hearings on Second Supplemental Appropriation Bill (1956),* pp. 474, 473.

4. The 1974 CIA report, initially secret, was published in 1976. Central Intelligence Agency, "Potential Implications of Trends in World Population, Food Production, and Climate," OPR-401, Aug. 1974, published as Appendix II to Impact Team, *The Weather Conspiracy: The Coming of the New Ice Age,* New York: Ballantine Books, 1977. For quote, see p. 200. News of the report was first published in the *New York Times,* May 1, 1976, p. 2.

5. Director-General of the United Kingdom Meteorological Office B. J. Mason, speaking mainly about aerosols and ozone, admitted in a lecture that greenhouse warming could become significant in 50-100 years. B. J. Mason, "Has the Weather Gone Mad?" *The New Republic,* July 30, 1977, pp. 21-23.

6. Hansen's 1988 prediction was based on an analysis presented in J. Hansen, I. Fung, A. Lacis, D. Rind, S. Lebedeff, R. Ruedy, G. Russell, and P. Stone, "Global Climate Changes as Forecast by Goddard Institute for Space Studies Three-Dimensional Model," *Journal of Geophysical Research*, Vol. 93, 1988, pp. 9341-9364.

7. Philip Shabecoff, "Draft Report on Global Warming Foresees Environmental Havoc in U.S.," *New York Times*, October 20, 1988.

8. T. Wilson, "Global Climate, World Politics and National Security," in V. Nanda, ed., *World Climate Change: The Role of International Law and Institutions*, Boulder, CO: Westview Press, 1983, pp. 71-77. See Wilson's article for more on the history of "national security" framing.

9. Quote from the IPCC. See J. T. Houghton *et al.*, eds., *Climate Change 1995: The Science of Climate Change*, Cambridge, England: Cambridge University Press, 1996.

10. Peter Schwartz and Douglas Randall, "An Abrupt Climate Change Scenario and Its Implications for United States National Security," Emeryville, CA: Global Business Network, 2003, available from *www.environmentaldefense.org/documents/3566_AbruptClimateChange.pdf.*

11. Mark Townsend and Paul Harris, "Now the Pentagon Tells Bush: Climate Change Will Destroy Us," *The Observer*, February 22, 2004. The Science Advisor to UK Prime Minister Tony Blair, Sir David King, called climate change "the greatest threat facing mankind" and "worse than terrorism."

12. Conservative consensus reports on expected impacts may be found at *www.ipcc.ch.* See also Nicholas Stern, *The Economics of Climate Change: The Stern Review*, Cambridge, England: Cambridge University Press, 2006 and HM Treasury, available from *www.hm-treasury.gov.uk/independent_reviews/stern_review_economics_climate_change/stern_review_report.cfm.* For scientifically well-grounded scenarios ranging up to a plausible worst-case beyond the IPCC and Stern reports, see Mark Lynas, *Six Degrees: Our Future on a Hotter Planet*, London: Fourth Estate, 2007. (Not yet published in the United States, available from *amazon.co.uk.*)

13. To learn more on the history, see *www.aip.org/history/climate*. See also Spencer R. Weart, *The Discovery of Global Warming*, Cambridge, MA: Harvard University Press, 2003.

The Science of Climate Change

Robert W. Corell
The Heinz Center

For the past year or two, many of us within the scientific community have been commenting that climate change is no longer just an environmental issue. It is now an economic and national security issue. It is good, therefore, to see so many of you gathered here to explore this issue in depth. The purpose of this presentation is to provide some basic insights into the science of climate change so that the reader can better gauge what we do and do not know at this point in time.

Let me start by elaborating a little further on some of the points made so well by Spencer Weart. First of all, a bit more should be said to underline the significance of the Vostok record. This dates back 650,000 years. It provides us with information about temperatures and CO^2 concentrations over the past 400,000 years. It is important to note that, at no time in the course of this 650,000 year period, did the level of CO^2 in the atmosphere reach the level at which it now stands. So, obviously, we are experiencing a slightly new condition.

Second, scientists have an increasing ability to measure climate change. Dr. Weart mentioned the use of ice cores. Why can we use these? When slowly deposited snow freezes, it freezes into tiny hollow spheres which capture the atmosphere as it was in the first few years of the snowfall. The air bubbles in ice contain samples of this past atmosphere. They contain CO^2 and methane and nitrous oxide and all the other gases. We have sufficiently good techniques to permit us to peel-out and look at this historic record of climate change. How, one might ask, do you measure temperature? It turns out that temperature is nothing more than the ratio of two isotopes of oxygen. These isotopes are to be found in the spheres, and so you can measure, quite accurately, what temperature it was at the time that the snow fell. So, the whole paleo record, the historic record, is well-captured by our ice cores. It is also well-captured in tree rings, coral reef cores, and sediments in the deep ocean. Thanks to increasing sensitivity of our instrumentation, particularly over the last generation, we can reconstruct temperature changes over time with increasing confidence.

After we came out of the last ice age—which ended some 20,000 years ago, as you may recall—we entered a period which I call the Anthropogenic Sweet-Spot. It lasted some 10,000 years. This period witnessed the birth of agriculture, the medieval warm period which caused the Vikings to go trucking off to Iceland and Greenland and ultimately even to North America, and a little ice age between the 15th and 18th century. All those changes took place in a temperature band of less than 1 degree centigrade. All the things we associate with evolution, from the birth of the concept of a village and the concept of a state, to lighting and heat in our homes, took place in this 10,000 year period.

We have now left this period behind us. The scientific evidence, in my view, moreover, permits us to predict with confidence that, at the current rate, we are likely to face a two to three degree warming period in this coming century. Hansen's research reveals that we have now come out of the Anthropogenic Sweet-Spot.[1] We did so (and I think most of the scientific community would agree with this statement) during the period in which there was a lot of sulphur and other pollutants in the atmosphere. These caused the temperature to level off before it went screaming northward again, and we kept getting higher and higher temperatures. There is a great deal of variation in temperature from month-to-month, season-to-season, and even year-to-year. The computer models tell us to expect such variability, and indeed variability is likely to grow as the temperature of the planet warms.

The Evidence from the Ocean.

The ocean is also very important to scientists attempting to understand climate change. Why is that so? Bear in mind that most of the water (97 percent) on earth is to be found in the ocean. Of all evaporative processes, 86 percent occur across the surface of the ocean. Oceans receive 78 percent of planetary precipitation. The oceans are the thermodynamic flywheel—they slow down the rate at which things happen, but they also give it momentum over time. Enough energy is stored in the ocean to make the planet even a bit warmer than it is now. And if you ask where the heat goes, that's where it goes. Any excess warmth that comes into this planet and is not reradiated out ends up in the ocean. So all the stuff we talk about is inside 10 percent. So the ocean really is the central game player.

Research undertaken by Tim Barnett gives you some idea of the role that heat is playing in the ocean.[2] Suppose we assemble temperature data from every available source (ships, satellites, etc.) for the last 40 years and plot it in all six oceans on a graph as red dots. Suppose we find that, in some places (e.g., the northern Indian Ocean), the warming does not extend nearly as far south as it does in, say, the north Atlantic. We then say to the computer: Well, let's back up 40 years and let's not allow humans to introduce their contribution to warming.[3] So the anthropogenic inputs are removed. Then we ask the computer to identify the statistical band within which the temperature is likely to be (blue dots). Then we run the computer again. We put in the equation, and the computer says that projections ought to be in the green. And then we ask the question for various U.S. computer runs: "For all these six oceans, what is the correlation coefficient between the actual measurements and the computer projections?" And it turns out they are all at 95 percent. We are becoming, in short, increasingly confident as time goes on. The computer models don't give us the fine details—they can't tell you, for example, what's happening here at Chapel Hill—but they can tell you very well what's happening on a global scale.

The Evidence from Icebergs.

The icebergs and glaciers of Greenland also help us to understand the process of global warming. In this country, floating icebergs extend some 700 meters below the surface. Greenland is also filled with glaciers which are really wonderful, magical things. Over the last decade our understanding of why they behave the way they do has changed radically. For example, at one point it was theorized that surface water did not make its way to the bottom. A research team from the University of Kansas has recently developed radar capable of penetrating the entire ice sheet, penetrating some 12,000 feet—something that 10 years ago we would have thought impossible and which, one imagines, the military must find interesting. We now find that there are lots of puddles of water below the ice that are lubricating these ice sheets.

In Greenland, ice is melting incredibly rapidly. Since 1979 the surface area has been reduced by about 30 percent. There is a place in Greenland where the face is coming off the ice sheet at a rate of 15 kilometers a year. If you stand in front of it for over an hour, you can actually see it move a couple of meters. So things are happening very rapidly in Greenland. As the Intergovernmental Panel on Climate Change (IPCC) says, both the Greenland and the Antarctic ice sheets, are reducing *en masse*. There is going to be some buildup due to snow fall. But, there is a net loss. The surface of melt is increasing about 1 percent per year, so we see Greenland as one of the hotspots—one of the places to watch, key our eye on. But why are they melting so rapidly? Well, if the ice sheets are percolating, they are taking the water down. If the water pooled on the surface, the energy exchange has to evaporate the water. That takes seven times as much energy as it does to melt ice. So if the water has disappeared and all you see is the surface ice, it takes one-seventh of the energy, so we accelerate the process of melting; so there is a dynamic feedback mechanism that is causing these glaciers to both melt faster and flow faster.

What are the projected temperatures likely to be in the high Arctic? As Dr. Weart noted, everyone knows for a fact that, no matter what you do, the Polar Regions are going to warm more rapidly, and the Arctic is going to warm more rapidly than Antarctica. Why is this so? The explanation is really simple. Take an ice drill in the Arctic and drill down two meters, and you're in water. So it's 0 degrees Centigrade. Go to Antarctica, you've got to drill 12,000 feet in ice. So, it's the difference between you putting just a tiny ice piece in your cocktail glass tonight versus filling it full of ice. The Antarctic mass is going to stay colder longer, and the high Artic is going to warm much more rapidly. There are other factors involved, but this is probably the dominant one. There are places which we expect to experience 8 to 10 degrees of warming from the 1990s to the 2090s.

What happened to Arctic Sea Ice in 2005 and 2006? The melt rate of the sea ice is pretty rapid. And though the rate has declined a bit in the last year or so, we are still talking about very large amounts of ice. What of thickness? Submarine data tells us that the ice was 40 percent thinner during this period than it had been previously. This is true, but one must bear in mind that submarines only go places where it has become 40 percent thinner. When you do the analysis, there is probably an 18 to 20 percent reduction in the thickness. This is a lot of fresh water. Dumping a lot of fresh water into the system also, incidentally, has some consequences as well.

Another interesting development to consider is the opening of new sea routes—notably the northern route and the Northwest Passage. Russia has always wanted to be a maritime nation. It is coming closer to that now. In fact, the sea route is 40 percent closer to the two major markets of Europe and the Far East. Two things have become clear since the last IPCC meeting. One, the opening is going to be on the Russian side. Two, it is likely to give Americans close to a half a year of opening within the next generation. The Arctic Climate Impact Assessment (ACIA) models project that the current navigation season of 20-30 days per year will increase to 3-6 months/yr by 2080, with one model indicating an ice-free summer by 2040.

This will have significant repercussions. All kinds of issues of access will have to be worked out. Seaward claims will be made. At the present moment, Canadians and Russians both lay claim to territory all the way to the Pole. Even the Americans, who generally adhere to the law of the sea, get quite heated over this particular issue. Boundary disputes are also likely to occur. There is already a boundary dispute between every Arctic nation. We are going to have to solve these problems and figure out the correct means to do so. Is the law of the sea going to be applicable? Or will there be some other forum in which to resolve these conflicts?

So what happened in the lower Arctic? We see the same picture. A lot of things are happening. In some places, where there is no longer an ice presence, the sea is eating away at the coastline and will gobble up little villages. In other places, where there is permafrost warming, the local fauna suffers. The Hudson Bay is almost empty of ice, and polar bears are well on their way to extinction now—they eat primarily seal and cannot hunt seals if there is no ice.

Intergovernmental Panel on Climate Change.

Let me finish by saying a few things about the key findings of the IPCC. These were released in February 2007.

First, the Panel concluded that "Global atmospheric concentrations of carbon dioxide, methane, and nitrous oxide have increased markedly as a result of human activities since 1750 and now far exceed pre-industrial values. The global increases in carbon dioxide concentration are due primarily to fossil fuel use and land-use change, while those of methane and nitrous oxide are primarily due to agriculture." The documentation is solid. Over the course of a 10,000 year period, things remained pretty stable. The last 100-150 years witnessed a pretty dramatic change.

Second, "Warming of the climate system is unequivocal, as is now evident from observations of increases in global average air and ocean temperatures, widespread melting of snow and ice, and rising global mean sea level." Note the use by the Panel of words like "unequivocal" which means 90 percent certain or better.

Third, "Most of the observed increase in globally averaged temperatures since the mid-20th century is *very likely* due to the observed increase in anthropogenic greenhouse gas concentrations." This also means that there is a 90 percent likelihood. The Report documented several long-term changes in climate: "The global average temperature trend over 1906–2005 is 0.74°C (1.3°F), increasing to 0.2°C (0.36°F) per decade over the last 3 decades; Global average sea level rose 0.17 meters (6.7 inches) over the 20th century; Mountain glaciers and snow cover have declined on average in both hemispheres."

One interesting thing to observe is that the IPCC projections for sea level rises which came out a few weeks ago were quite low. That is because they decided not to include glacial ice-sheet melt in their model because they did not have sufficient confidence in those predictions. The IPCC has a very calming effect. They ask, what can we scientists agree upon? By the way, most of the sea level ice comes from thermal expansion into the water and not from the sea ice—not sea ice but glacial ice sheets. But this century we're going to see an increasing contribution from Greenland and ice sheets around the world. Virtually all (probably 98 percent) glaciers are now losing mass. The IPCC also made clear that predictions are scenario dependent. For example, if we consider how far sea level might have risen by the end of the 21st century, we can find low and high projections. The low projection is: 0.28 meters (11+ inches). The high projection is: 0.39 meters (15 + inches). Recent literature projects a rise of about 1 meter.

What might happen in the United States? Studies suggest we will see dramatic shifts in temperature regimes, though scenarios vary. Projections must take into consideration a variety of factors to assess how a coastline might be affected: tidal range, wave height, coastal slope, coastal shoreline change rates, geomorphology, historical rates of relative sea change. Virtually all the models we now run say that the Southwest is really going to get hit by droughts of a pretty dramatic nature. These droughts may last, not just a few years, but many decades. Analysts think it might take as long as a century to recover. Beyond that, no one is willing to talk. We're going to see shifts in the type of vegetation, and entire ecosystems will change.

What about time scales? If we're really aggressive and say, "Okay, we're going to bring our greenhouse gases down over this next century," what happens? First of all, it's going to take several hundred years to stabilize CO^2 because its resident time in the atmosphere is 120 years. So it could take some time for this supertanker called climate change to level off. This is a new world—a world that is two to three degrees warmer will look a lot different from the world we are familiar with. Everything we are used to emerged during the human "sweet spot" of relative climatic stability of which I spoke earlier. At this higher temperature, ice is going to continue to melt, and the oceans are going to continue to experience thermal expansion. There is a real lack of symmetry between how long it took to set these changes in motion, and how long it will take to stop/reverse the problem. Even if temperatures are stabilized within the next hundred years, it will take centuries to millennia for sea level rises to stabilize.

How are we going to come to grips with this problem? There are solutions. We can take one step at a time: increase fuel standards, reduce vehicle use, increase energy efficiency, etc. If you do that, you will start undoing some of the damage. Eventually over the course of the next 100 years, you may be able to stabilize matters.

In the final analysis, remember this is where we are. We are living on a tiny little planet. It is the only place we have. We need to take care of it.

ENDNOTES - Corell

1. Jim Hansen is Director of the NASA Goddard Institute for Space Studies and Adjunct Professor of Earth and Environmental Sciences at Columbia University's Earth Institute. During his famous congressional testimony given in the hot summer of 1988, he showed GISS model projections of continued global warming assuming further increases in human produced greenhouse gases.

2. Tim Barnett is a research marine geophysicist in the Climate Research Division of Scripps Institution of Oceanography, University of California, San Diego. His research focuses on the physics of climate change and long-range climate forecasting. See, for example, "Detection of Anthropogenic Climate Change in the World's Oceans," *Science*, Vol. 292, 2001, pp. 270-274.

3. The nice thing about the past 40 years is that we have a pretty good idea about the CO^2 and methane and other things.

A Threat Assessment

Richard A. Matthew
University of California, Irvine

Introduction.

The purpose of this paper is to put climate change into the context of global politics, talk a little bit about the work that has already been done linking environmental change and security around the world, and then look at ways we might think of linking climate change to security. The geographic focus will be on South Asia.

The previous two panelists have already described in detail the science of climate change, so I will not reiterate any of this, but will turn immediately to how climate change interacts with other global forces. I also will raise some questions about what we mean when we use the term security. I do, however, want to make two points about the science. First, we should bear in mind that the **pace** of change (whether global climate change is abrupt or gradual) will affect security. Second, while scientific models capture broad trends very well and help us to imagine the future, the precise impacts of climate change are likely to vary enormously from place to place, which means, of course, that the security effects will vary. We know that we should expect such consequences as water scarcity, desertification, and sea level rise. We know that changes are going to take place that will affect food production and microbial activity. We do know that all these elements could become security concerns, but we cannot predict how and where with any precision.[1]

There is a good reason for this, and it relates to the broader point I want to make, which is that climate change is taking place in the context of a lot of other global changes. In the next few decades, we are probably going to be living in a relatively new world, in which humans experience things not experienced before. It is going to be a new world in the sense that its climate will be different. But it is also going to be a new world in a lot of other ways. The natural environment is undergoing multiple forms of severe stress due to land and energy use, and significant social changes also are taking place throughout the planet.

For example, an important demographic change is taking place. We are soon going to be living in one the oldest societies known to humankind. Some of the planet's societies are going to move towards an average age of 50 during the course of our lifetimes. This has all sorts of implications for things like health care needs and retirement planning, but it also is likely to interact with climate change. Specifically, we know that things like heat waves take a higher toll on elderly people.

What else is happening? We know that we are in a world in which a vast and unprecedented informal economy has grown up alongside trade liberalization. This informal economy has received scholarly and policy attention because it has become home to all sorts of criminal activity, as well as a rich menu of poverty alleviation initiatives. But the unregulated character and sheer size of the informal economy also has implications for fuel wood and other energy sources.

We know that global terrorism, or transnational networks of terrorists with global agendas, has become an increasing problem in the last decade and is likely to continue to be a grave threat. We can imagine scenarios in which energy supplies are attacked during heat waves, causing considerable problems to us.

Another notable global change is the extent of democratization that has taken place since the end of the Cold War, something that has been in many ways a tremendously positive development. I just got back yesterday from a meeting in Stockholm with a group of people from Nepal who are at the forefront of the democratic change taking place there. The excitement and the enthusiasm these people express are tremendously affirming. However, the expectations that they have for quick and permanent social gains are remarkably unrealistic in a lot of ways. It is not easy to quickly provide extensive public goods, and opportunities to acquire private goods, in places that are characterized by enormous inequality. Indeed, democratic efforts have stalled twice in the past in Nepal. The same thing has been true for a number of other countries around the world trying to make this change. Nothing is certain, but climate change could conceivably deepen inequalities in ways that make political reform even more daunting.

One of the most widely observed forms of global change has to do with the pace and extent of technological innovations and diffusions that give people worldwide access to information and an unprecedented ability to learn about and become involved with things that are taking place far from where they live and work. I was talking to a group of epidemiologists at The University of California-Irvine, and they were describing how quickly the world is sharing scientific information these days. Real time global scientific collaboration meant that scientists were able to understand the characteristics of Severe Acute Respiratory Syndrome (SARS) far more quickly than they did Human Immunodeficiency Virus (HIV)/ Acquired Immune Deficiency Syndrome (AIDS). Of course we also know that there are dangers that have developed which are inherent in the globe spanning information and communication technologies, as we have no effective way of denying terrorists and criminals access to them.

Another positive development around the world is the steady institutionalization of the empowerment of women. This, too, has implications that we can scarcely anticipate.

In short, a lot of things that are happening in the world are changing people's values, practices, institutions, and beliefs. We do not know exactly what they will mean for complex processes like global climate change. But clearly, climate change does not operate in a vacuum. Rather, it competes and interacts with a lot of other important global forces.

Defining National Security.

At a conference dedicated to studying the national security implications of global climate change we need to think about what we mean by national security. Political scientists who are focused on security studies and international relations tend to look to things like territorial integrity, critical infrastructure, national identity, protection of people, a government's ability to govern, and our military's readiness when they seek to define national security. But in a world that is changing dramatically, it may be that these things, so long a part of security studies, need to be reconsidered. For example, compare the notion of territorial integrity to the idea of a state's ecological footprint, that is to say, the amount of resources that a country uses to maintain itself. For much of the world, the latter is far larger than the former. The United States, Canada, Australia, and a number of other countries have such an abundance of resources that we and they could probably maintain a very healthy lifestyle for some time without drawing on resources from the rest of the world. In the case of other countries—Japan, Israel, and Switzerland, for example—territoriality does not really define what they need to be secure in the sense of being able to maintain their standard of living.

Critical infrastructure no longer stops at national borders. Technologies have integrated our information and communication systems into global systems, and while we depend on these, they have a life of their own.

National identity is being changed. Here in the United States we can probably cope—our country is accustomed to tremendous diversity and to people engaged in complex relationships with the rest of the world. But many countries must deal with a sudden influx of large numbers of people who pose a challenge to their traditional identity.

In a recent book, Princeton professor Anne-Marie Slaughter argues that our government is already becoming transnational. She says, in effect, that we can no longer maintain the sort of image of territorial and political autonomy that was so popular during the 19th and 20th centuries. In fact, if we were to cut off our government from the rest of the world, we would reduce its capacity in a significant way.[2]

These challenges to conventional ways of thinking about security are significant. They may be more advanced elsewhere, and we may be coming to terms with them later than, say, the countries that have formed the European Union and really grasp the logics of interdependence. Very likely, because of the somewhat erratic character of global change, national security will not mean the same things to all people for a very long time—if ever. I think that today our cherished and historically grounded definitions are not as persuasive to everybody as they were during the 20th century.

Climate Change and Security.

Both the planet's climate and its understandings of security are in flux. Fortunately, when we try to think of how to link climate change to security, we are not starting with an entirely blank sheet of paper. For the past 20-30 years, researchers have been working to link environmental change and security in a variety of ways.[3] Now some people who study security argue that this linkage does not have a lot of explanatory power or add much that is new. They believe that the existing set of theories about war, conflict, and threat can accommodate things like migrations due to desertification or competition over oil, which trigger processes no different than the familiar competitions over state power or market share.

The environmental security scholars do, however, have some important and novel insights. For example, they appreciate the security significance of population growth.[4] Not everyone agrees on statistical estimates, but world population is expected to grow by a couple of billion during the next 50 to 100 years. It is well-established that we are already overusing the bio-productive capacity—the renewable food and energy—of the planet. Consider that three billion people are today living in conditions of dire poverty and that another two or three billion people are going to be added into this category. Consider further that we are currently using some 2.3 hectares of bioproductivity per person to maintain the lifestyle that we have globally today, and that we probably do not have more than about one and a half hectares of bioproductivity to give to each person on the planet. In other words, we are using up resources like fish and forests faster than they can be replenished. Indeed, influential scientists like E. O. Wilson, the Harvard biologist, are even more pessimistic about the relationship between renewable resources and consumption rates than what I am suggesting here.[5] This poses a very serious problem. It means that there are a whole lot of people who have little, if any, prospects of ever pulling themselves out of dire poverty and the imperative of basic survival, no matter how unsustainable this may be. It would be remarkable if technology could close this gap between bioproductivity and consumption in the space of the next 50 or 100 years. Environmental security scholars looking at the social implications of this type of scarcity are doing useful work which does, I think, have enormous implications for climate change, which is going to place further pressure on many renewable natural resources.

Another line of scholarly research that has been pursued examines the relationship between natural resource abundance, greed, and violent conflict. While the widespread and violent competition over things like diamonds, gold, and oil throughout the world does not add much in terms of theory to our understandings of security, we should be aware that competition over resources like oil could increase as the world warms up or as other types of changes take place and the demand for energy grows.

According to last year's *Human Security Report* put out by the University of British Columbia, the world is overall becoming a more peaceful, more cooperative, and less violent place. This is obviously, on the whole, encouraging. But although there have been great gains in terms of reducing the number of people killed in war and displaced by war, there are, nonetheless, many seemingly intractable areas of extreme violence. These areas do largely coincide with areas where one finds conditions of scarcity or violent competition for control of a natural resource like oil or gold.

So the big question is what will happen as we add two billion more people to a planet which is not producing enough for the six billion people already in existence? Will the additional stress of climate change create the kind of conditions that make it difficult, or even impossible, to find rational ways to meet the needs of the world? Will collaboration continue, or are we going to see increased competition for pieces of a pie which will be getting smaller and smaller for a lot of people? Evidence from the Sudan and Rwanda and throughout South Asia suggests that there is a very tight set of connections between forms of environmental stress, lack of access to credit, confusion over property rights, and poverty and inequality.[6] There is also a strong relationship between these conditions and the existence of violent conflict and other forms of human insecurity.[7] I think the environmental security literature has introduced into the security studies field some ideas that are significant and worth taking seriously.

Against this background, climate change and security can be linked in a number of ways. Where climate changes abruptly, security problems will be immediate and extensive and perhaps even existential. We can easily envision threats on this scale in Bangladesh or other poor low-lying countries, but even here a significant number of Americans would be affected by a sudden barrage of massive flooding, Katrina-sized hurricanes, and tropical disease epidemics—perhaps enough to make climate change a national security issue. Another possible threat that we should take seriously is that of the gradual erosion of American power as endless demands are placed on it due to abrupt changes elsewhere. These are likely to arise as we face humanitarian disasters, as drought intensifies throughout Africa, and as South Asia collapses into conflict over things like fresh water. The greater our sense of interdependence, the greater our sense that national security depends on the welfare of things beyond our borders, and the more likely it is that the climate change will be a real security threat. This poses a big problem today. To what extent should we intervene to assist abroad? When should we use our resources and when should we show restraint? It is going to be difficult to make these decisions. We are playing with a lot of uncertainty. We do not know how other actors in the world will behave.

Gradual climate change, by its very nature, creates more opportunities for effective adaptation and mitigation measures, but I want to raise one concern—complacency. Like many things in life, the sooner we invest, the cheaper it will be. I think as a nation we are ignoring a lot of compelling data about what is happening because so far the changes inside our country have been quite manageable. But think of the models of the first two speakers. Now is the time to work hard to reduce energy use, increase efficiency, implement green design, revise our education and research programs, adopt alternative energy forms, and reduce consumption.

Further complicating matters, we also know that there will be winners and losers as the world's climate changes. Not everyone will experience the same kind of problems, and some areas will find the changes conducive to human settlement and increased agricultural output and so on. But overall, the expected downside massively outweighs any predicted upside. The menu of likely threats includes severe weather events, changes in the food supply, massive flooding, and dramatic changes in microbial activity that will lead to the spread of infectious disease. Indeed, many analysts believe that we are very close to a global pandemic. They anticipate a transfer of disease from the animal kingdom to the human kingdom that will be highly virulent. A lot of these transfers have taken place in the past 3 decades because environmental conditions are changing and because people are being forced into marginal environments where they come into close contact with pathogens with which they have not had any contact in the past.

South Asia.

At this point, let me put a concrete face on all of this by looking at a specific place. Let me examine what might happen in South Asia. A quarter of the world lives there. It is a region already facing water scarcity. Much of the fresh water comes from the mountain regions. If the glaciers continue to melt and snow patterns continue to change as we now see happening, people may, for a couple of decades, think that there is a lot of fresh water to be had. That is, there will be a temporary increase in many areas. But this will suddenly and quite abruptly change, and the people there will soon find that there is, in fact, a real scarcity of fresh water in an area that has one and a half billion people and in which there are two nuclear powers.

Now there is disagreement in the academic literature over the extent to which competition over water leads to violent conflict rather than being worked out institutionally. But whatever the trends of the past, there is general agreement that climate change could create an entirely new type of water politics. South Asia is an area of considerable concern in this regard. If we look at water-related conflict in South Asia, there have been 10 incidents of violent conflict in the past 7 years. In the 53 years before, there were only three incidents. In other words, water-related violent conflict has increased 24-fold in the past 7 years. Fresh water is clearly the cause of grave concern throughout the subcontinent these days. Nepal, Pakistan, and Bangladesh worry that India is using its enormous power to negotiate a series of bilateral agreements that may not be on their best interests. So these countries are already very nervous about their prospects for fresh water over the next 20, 30, or 40 years, but they are not sure what to do.

They are aware, however, that their fresh water supply could collapse dramatically, and they are concerned about how this will play out. Let me give you one example from my own research that I think is somewhat illustrative of what might lie in the not too distant future. Nepal, of course, is a fairly small country, 30 million people, but it is an extremely poor country, a country with a very low literacy rate and low rates in things like the United Nations (UN) Human Development Index, where it ranks at the very bottom of the planet's 192 countries or close to the bottom. It is a country that has experimented with democracy. It has also experienced 10 years of extreme civil conflict that has driven the government to invest less and less in education and fresh water and sanitation and more and more in security. Nepal has a population that is very youthful, growing very rapidly, concentrated in a small number of areas. In short, it has all the conditions for violent conflict. It is transforming into a democracy, it is extremely impoverished, there is a high population density, and the big issue today is land. During 10 years of insurgency, the Maoists promised to redistribute land once they came into power, but since they joined the parliamentary system in November, they have backed away from this promise in two ways. Now they are saying, "Well, we will not redistribute any land of less than 10 hectares because that would be costly and disruptive, and, as for the formula for the rest, we have no idea what it should be, because we do not want to throw our economy into chaos by scaring away our local expertise and foreign investors."

So right now the politics of Nepal is defined by a tremendous expectation for land reform and very little clue of how to reform land ownership and access. What people are also starting to recognize is that land reform will need to be somehow integrated with access to fresh water. But fresh water is something that Nepal has been losing. It sees itself as a water rich country, but it has been losing a considerable amount to India. For example, in the 1950s people were encouraged to move to a wetland area of the Koshi River system to take pressure off the Kathmandu Valley. Half of this wetland was then leased to India because India has tremendous thirst for fresh water and needed it for irrigation in the north. Then the other half was turned into a protected site because the people were destroying it. What happened to the people? Well, the people, feeling uncompensated, appear to have been mobilized by the Maoists, who promised to return the wetland to them as soon as they were in power. Now you have a situation in which millions of people are waiting to recover or gain access to water and land that they believe is rightfully theirs, and the Maoists have no idea how they are going to satisfy this expectation—

one they cultivated. Meanwhile, Maoists have risen in power significantly in India in the past couple of years. These Naxalites are calling the Nepali Maoists soft because they are not dealing with the water and land issues that they promised they would deal with. This is a sort of pattern we are going to see more and more of. Ultimately there is no easy solution to the land and water issues in South Asia. Climate changes are likely to make these problems even more difficult to solve than in the past.

In conclusion, we face a gap between what we need to do to ensure security and what we are actually doing. Technology has moved us into a new world where we have new needs. Our old institutions are not adequate to meet these needs. The question is, are we going to be able to develop new institutions? Or are we going to try to use our existing ones? Time does not permit us to elaborate on how we might close this gap but clearly research on institutional reform and cooperation are key elements.

ENDNOTES - Matthew

1. The social effects of climate change, and of many other forms of environmental change, are varied and hard to predict. This is in large measure because such a complex array of interactive variables and nonlinear relationships shape social outcomes. For a further discussion of this, see Richard Matthew and George Shambaugh, "Sex, Drugs and Heavy Metal: Transnational Threats and National Vulnerabilities," *Security Dialogue*, Vol. 29, Summer 1998, pp. 163-175.

2. Anne-Marie Slaughter, *A New World Order*, Princeton, NJ: Princeton University Press, 2004.

3. For example, see Peter Gleick, "The Implications of Global Climate Changes for International Security," *Climate Change*, Vol. 15, 1989, pp. 303-325; Steve Lonergan, *Global Environmental Change*

and Human Security Science Plan, IHDP Report 11, Bonn, Germany: IHDP, 1999; Richard Matthew, Ted Gaulin, and Bryan McDonald, "Elusive Quest: Linking Environmental Change and Conflict," *Canadian Journal of Political Science*, Vol. 36, 2003, pp. 231-256; Jeffrey Sachs, "Climate Change and War," 2005, available *from www.tompaine.com/print/climate_change_and_war.php*, accessed March 11, 2005.

4. For example, see Paul R. Ehrlich, *The Population Bomb*, New York: Ballantine, 1968.

5. Edward O. Wilson, *Diversity of Life*, Cambridge, MA: Harvard University Press, 1992.

6. On Asia, see, for example, Richard Matthew, "Sustainable Livelihoods, Environmental Security, and Conflict Mitigation: Four Cases in South Asia," IUCN Poverty, Equity and Rights in Conservation Working Paper Series, 2005, available from *www.iucn.org/themes/spg/Files/IUED/Case%20Study%20South%20Asia.pdf*. On the Sudan, see, for example, The IISD project on *Climate Change, Resources and Conflict: Understanding the Links between Environment and Security in Sudan*. R. Matthew is a member of the project team. Further information is available from *www.iisd.org/natres/security*.

7. For example, see Robert Kaplan, "The Coming Anarchy: How Scarcity, Crime, Overpopulation, Tribalism, and Disease Are Rapidly Destroying the Social Fabric of Our Planet," 1994, available from *theatlantic.com/politics/foreign/anarchy.htm*; Thomas Homer-Dixon, *Environment, Scarcity and Violence*, Princeton, NJ: Princeton University Press, 1999; Michael Klare, *Resource Wars: The New Landscape of Global Conflict*, New York: Henry Holt and Company, 2001.

Commentator

James A. Rotenberg
University of North Carolina, Wilmington

At the G8 conference back in 2005, British Prime Minister Tony Blair referred to Climate Change as "Probably long term, the single most important issue we face as a global community. . . ." The Intergovernmental Panel on Climate Change (IPCC) Climate Change Report which came out earlier this year made the same point. On page 5, it says that the warming of climate is unequivocal. We are now about 90 percent confident of the accuracy of our data. We not only have historical data and more recent climate change models, but we also are more certain in our understanding of how climate change might affect security. We probably should mention that not everyone agrees—it has been said that global warming is the "greatest hoax ever perpetrated on the American people."

Let us briefly put this issue into a broader context. I myself am a tropical environmental ecologist. I study birds as environmental indicators of change. Global warming affects human populations, as we have heard. It also affects other biological organisms. A recent study looked at the range of occurrence of various different birds of about 35 neotropical migratory species that fly south for the winter and come back to North America to breed in the spring and summer. The breeding range of seven of these have already shifted significantly in the past 25 years on an average of more than 65 miles, likely due to climate change. What does that mean or why should we even care? In fact, these birds can serve as the proverbial canary in a coal mine. They can provide us with significant information about the availability of resources. Another study done by Jeff Price for the American Bird Conservatory used a climate model to model bird activity and the bird activity to look at future rain distribution for North America. The model predicted definite net changes for birds all across America—with some birds moving to different areas in response to climate, and other species being lost altogether. E. O. Wilson (preeminent scholar of Ecology, and author of books such as *Biodiversity* and *The Future of Life*) recently visited UNC-Wilmington. I asked him, "What is the greatest issue facing us right now, Dr. Wilson?" Immediately, he responded with "climate change," and in particular he went on to say that climate change and global warming will alter our resources and resource needs for the future.

I'm sure most of you have heard of Rachel Carson's *Silent Spring*. Why "silent?" Well, because the birds were not singing anymore. The book, published in the 1960s, was about pesticide use and its negative effects on birds. A few decades later, John Terborgh wrote a book on bird conservation, entitled *Where Have All the Birds Gone? Essays on the Biology and Conservation of Birds That Migrate to the American Tropics*, which addressed bird population declines due to habitat destruction in the topics. Will the pivotal book of the next decade be on *Birds and Climate Change*?

Let me conclude by posing several questions. First, how will environmental change, in particular, resource needs, such as water, food, and energy, shape our world in the future? To what extent will security concerns affect our thinking on the environment? This first panel has made me realize that I do, in fact, address security dimensions in my global environmental class—I talk about such matters as the needs of the poor, the changing demographics of human populations, and how water scarcity may really change how we look at things. Second, what is the role of the military? Can they really combat climate change? And third, can we be a giving nation as well as a nation that sees to its own needs? Thank you.

Discussion

Q: Can global climate models predict **local** changes? If not, what can be done to improve our capability in this area?

Corell: Global models, which call for supercomputers, do not provide nuanced information about local changes. To get a finer scale, we have developed two techniques. The first entails dynamic downscaling, which is expensive and still calls for powerful computers. You nest a finer grid within the larger grid and connect the two. The second is called statistical downscaling. You use weather stations and databases to characterize what is going on in a region over a given period and then build a statistical relationship between that fine scale behavior and large scale computer models. It takes about 30 years to get the statistics to stabilize. The results are interesting: In Norway, for example, the computer model *Global One* says there is no temperature difference between north and south. When one is using downscaling, we get a five to one difference.

Q: You talk about the gradual erosion of American power that may occur as humanitarian disasters take place what do you mean?

Matthew: Our military capability is only so great. Last year there were 800 natural and human disasters around the world. If that number continues to increase as predicted, our resources will be strained.

Q: How high will sea levels rise in the course of this century?

Corell: Estimates differ. In its February 2 report based on the computer models, the IPCC suggested something on the order of a foot and half. Remember, that they cannot use any data that has not been published in peer reviewed journals. They also do not include material published in the last 18 months. Most scientists would argue that the ice sheet melts from Greenland and Antarctica, and from the glaciers of the land mass, will push that number up closer to a meter. I don't think any responsible persons are talking about a three, four, or five meter sea level rise during this century. Remember, though, that there will be differences—rather remarkable ones—around the globe, depending on the topography and other effects. The Chesapeake Bay, for example, is rising at twice the global rate. That is because it is a Helmholtz resonator which amplifies effects as tides move about. It also experiences subsidence. There was a glacier there that ended by the bay bridge. As the glacier formed, the elastic soil pushed a bow wave ahead of it. As it goes away, this wave subsides—that is what causes the 50 percent relative sea level rise. All this causes confusion. I would argue in favor of an estimate of about a meter rise in the general sea level this coming century.

Weart: I was at a meeting very recently with the Assessment Group looking at rapid ice sheet change. They noted that ice sheets are more sensitive than previously thought to temperature rise, so we may see more than this. And we may also see very rapid rises. A surge that would raise sea levels by a meter within a few decades would be very catastrophic for coastal populations.

Q: A professor at The Massachusetts Institute of Technolgy (MIT) spoke of a moderating effect that will keep the temperature rise to one or perhaps one and a half degrees over the next 100 years. Have you heard this explanation?

Corell: Yes. I have debated this with him in public and private. One of the beauties of science is that there is always someone to challenge the system. The lonely voice ought always to be encouraged but not necessarily believed as equal to the collective wisdom. We should remember that the IPCC arrives at its conclusions only after looking and judging a wide range of scientific papers and after a lot of debate and discussion. It also chooses its words very carefully. So when it says something is likely or very likely, that has statistical meaning.

Q: Can you speak to the likelihood of abrupt climate change?

Corell: This is something which we are still struggling to get a handle on. Most of the models are monotonic in their behavior. Our best estimates use what we might call the analog. This means we go back in time and try to find a case where the conditions were similar and then draw some conclusions about abruptness. At the moment we can speak with much less confidence about abrupt than long-term change. We do know, though that these things happen: We know, for example, that the thermal hyaline circulation was dramatically readjusted in the North Atlantic as we came out of the last ice age. At the Heinz Center, we are looking at threshold systems/tipping points. We can do a little better in ecological systems. Certain plant species are temperature sensitive, so changes in their growing area can tell us what is happening. Those tipping points are driven by temperature, and other tipping points are driven by phase change between ice and water. The methane release in the Arctic is going to be dramatically affected by the bog and permafrost melt. On the grand scale, I would say it is still pretty tough to give you some sense of what will happen—in a century or two we might see thermal haline circulation adjustment.

Weart: It is possible to frame climate change as an insurance problem. In the late 1990s reinsurance companies and others who handle events like massive hurricanes began to be concerned that climate change might increase the occurrence of such abrupt events. We could, I think, follow the insurance model in national security. When it comes to climate change, there are things that we are pretty sure will happen. We need to invest considerable resources into dealing with these. At the same time, there are things that probably won't happen but could. Given their potentially catastrophic nature, we must still make some provision for these.

Corell: I want to build on that. We need to put this whole problem in some sort of a risk management framework. The military really does understand how to evaluate risk. For us to govern our lives by the mean temperature would make no more sense than for military strategy to be dictated by the average of something happening. It makes sense to put what we are doing into some sort of risk probabilistic framework. I recently met with Pacific Gas and Electric. It took some of the senior managers a while to take to the notion that their company ought to be become the first ever to be fossil free. What made them see the value in what we were doing was when we asked them, how many of you ever have had your house burn down? No hands went up. How many of you have fire insurance? Every hand went up. Nesting climate change and these other factors that are affecting our future in this framework will go a long way towards giving us a much sounder ground on which to work.

CHAPTER 2

HUMAN SECURITY

How exactly climate change and conflict are linked remains an area of controversy. Scholars disagree as to how far resource shortages, the spread of disease, human migration on a massive scale, and other climate-generated factors might exacerbate existing conflicts. In Chapter 2, Dr. Erika Weinthal considers water-generated conflict, Dr. Andrew Price-Smith focuses on disease, and Dr. Timothy McKeown analyzes the complex way demographic shifts relate to conflict.

Water, Climate Change, and Human Security

Erika Weinthal
Duke University

I was asked to speak today about water, which is a very broad topic. Although most of my work on water has concentrated on the Middle East and Central Asia, events elsewhere highlight the tremendous disparity in freshwater resources worldwide. In particular, I am referring to the tremendous downpours and flooding in Europe over the last few years and, most recently, in Indonesia. Yet, elsewhere, the picture is very different. In the Middle East, shortages are increasingly the norm. Take the river Jordan. We think of it as the "Mighty" Jordan, but in the summer, it is a mere trickle. You cannot even bathe in it. Water scarcity, is, in fact, where I want to begin. I then want to tie this issue of scarcity to climate change, and consider how this combination has an impact upon both human security and conflict.

My first point is that when we think about water as

a source of conflict, water scarcity usually is the first thing that comes to mind.[1] There are several reasons why this is so. First of all, the global distribution of water is uneven. Some states have plenty of water—for example, there is a tremendous amount of fresh water in Lake Baikal in the Russian Federation, and Latin America has about 31 percent of global fresh water—while other parts of the world (e.g., Northern Africa and the Middle East) are poorly endowed.

Secondly, it is not just that water is unequally distributed, but also that most of the world's major sources of freshwater are shared between states, forcing states to negotiate with their neighbors to determine the appropriate distribution of resources. There has, nonetheless, been a greater potential for conflict among riparians since the end of the Cold War, largely owing to the breakup of the Soviet Union and Yugoslavia, which produced several new international river basins. Whereas in 1978, there were 214 shared international river basins, today there are 263.[2] Take the Aral Sea Basin. During the Soviet period, it was an internal river basin whereby Moscow would decide how to allocate the water among the five different Central Asian republics. When the Soviet Union collapsed, new territorial borders sprang up, and conflicts arose between the upstream and downstream riparians in Central Asia over how water would be utilized for both irrigation and hydroelectricity generation.[3] The Danube River basin, following the reconstitution of East Central Europe, is now shared by 19 countries, and here conflicts have transpired in the early 1990s over the Gabcikovo-Nagymaros dams.

The situation is no different if you look to other regions. In South Asia, the Ganges Brahmaputra is shared by six states and the Indus by five. In the

Middle East, the Jordan is shared by five. Finally, in Africa, there are a large number of river basins that are shared—the Congo, the Zambezi, the Okavango, the Volta, the Niger, and the Nile, which is shared by no less than 11 countries.

Water sharing in Africa is especially complicated because it is a region that has been ravaged by numerous internal conflicts over the last few decades. It is also a region that will likely be greatly affected by climate change—in particular, an increase in desertification along the Sub-Saharan zone. Thus, finding solutions to Africa's water crisis will also invariably entail dealing with the effects of climate change as well as other socio-political issues related to water resources distribution.

The region that is usually most associated with water related conflict is the Middle East. In a famous quote, Boutros Boutros-Ghali said: "The next war in the Middle East will be fought over water and not politics."[4] A number of people have suggested—not without good reason—that water will drive future political conflicts in the Middle East.

For example, how the waters of the Jordan River should be used and by whom has been very contentious for those riparians with a claim to its waters. That the issue is so highly contested is attested by its inclusion in the formal peace treaty between Jordan and Israel. Similarly, the Oslo Accords had a section on water and a section on the environment, as water is of vital economic importance to both Israelis and Palestinians. The Israelis and Palestinians agreed to negotiate a solution to share the mountain and coastal aquifers at the height of the Oslo Accords. This is unusual. In the aftermath of wars, there is usually a tendency to postpone dealing with really heated issues—and in the Middle East, water belongs to that category.[5] Thus, it is rare to find water (and other environmental issues) directly included in a peace treaty. In both these cases, however, peacemakers clearly recognized that a resolution to the issue of water sharing was essential for maintaining the peace and rebuilding economies and societies.

It is the **scarcity** of the resource, then, which is most closely associated in people's minds with conflict and human insecurity. And, to tie this talk back to climate change—the primary theme of the conference—we do know that changes in precipitation, temperature, and carbon dioxide levels will affect the supply of and demand for renewable water resources. However, scarcity is only one part of the picture. There are other things which we need to factor in when assessing the role played by water in conflict, including the capacity of state institutions to adapt, the quality (as opposed to simply the quantity) of water, and demographics.

The effects of climate change will vary because some states are more vulnerable and less able to adapt than others. Developing countries often do not have the institutional or technological capacity to plan for some of the changes that are most likely to come about. When we come up with models, it is important that we consider what kind of states we are looking at and think about whether or not we need to develop different planning strategies for different states—that is, the ability to implement our models might be contingent upon the particular institutional configuration within any one state.

Water quality, moreover, must also be taken into consideration. There are 1.1 billion people worldwide that lack access to clean water, and 2.6 billion people that lack access to sanitation.[6] Most of this population, again, is in Sub-Saharan Africa or in Asia. A society—even one likely to be dramatically affected by climate change—is not likely to give much thought to planning for this eventuality if most of its energies are focused on just trying to procure water to meet very basic human needs.

Let us return to a consideration of the Middle East. This region is very arid. Countries with an annual availability of less than 1,000 cubic meters per person are considered to be water-scarce. Countries which have less than 500 cubic meters annually are considered absolute scarce. The Gaza strip, for example, is very water stressed.[7] Each person has access to no more than about 320 cubic meters annually. The water resources—from the coastal aquifer—are shared between Israel and the Palestinian Authority. Here, water flows from Israel into the Gaza Strip—that is, towards the Mediterranean Sea. In the Gaza Strip, the water quality is extremely poor—the water salinity is very high. In addition to an overall water shortage, there is a significant shortage of safe drinking water. Add to this the fact that Gaza has one of the fastest growing populations in the world—about 4 percent growth rate per year. It is very common for individual women to have about seven children. Nearly 1.5 million Palestinians live on a very small piece of land, and the population is expected to reach 2.5 million in the next decade. Given the current consumption rates of water resources, the natural replenishment rate is far lower, which means that there is a growing water deficit and increasing magnitude of the water crisis.

Attention must also be paid to demographics. Population growth is going to exacerbate the effects of climate change especially in regions like the Gaza Strip. The net effect of climate change on water availability will be limited in some regions. In other regions, climate change, compounded by population growth, will actually have a much greater impact globally.[8] This is especially true of regions such as North Africa and the Middle East. It is estimated that in the year 2025 about 3 billion people might be living in water-stressed areas.

How will climate change affect the more long-term prospects of a post-conflict society? We tend to focus on the conflicts themselves and how to bring them to an end. We do not devote enough time to thinking about the next stage. What happens after a conflict ends? A recent finding suggests that many people will die following a civil war because they lack good drinking water and are forced to live in unsanitary conditions.[9] Far more people, in fact, die every year from poor water quality than they do in war. For example, 1.8 million children die from diarrhea and water-borne illnesses yearly.

People also migrate during and after conflicts. This not only leads to the spread of infectious diseases but it puts added stress on strained water resources. Consider, for example, the case of Iraq today. Thousands of people are moving into Jordan. It is one of the few places they can go. They have also been trying to get into Syria. Water is already scarce in these areas. In Jordan, current supplies barely provide enough water for its rapidly growing population. Thus, the government could face mounting pressure to develop more rapidly its fossil (i.e., nonsustainable) groundwater. Consider also the case of Bangladesh. Here, if the models are correct, monsoons will be more intense, and flooding will also be more intense in the future. In that case, we are likely to see an unprecedented scale of migration within Bangladesh and into neighboring countries such as India. Groups are likely to compete for scarce resources and/or poor quality resources—another situation ripe for conflict.

Before I conclude, I would like to return to the question of adaptability. Why are some countries more vulnerable to climate change than others? More particularly, why do some countries choose *not* to adapt? Sometimes it is because to do so would require reforming their economies, which would entail tremendous social and political costs. To understand this, one must understand something about agriculture and water use and bear in mind that approximately 80 percent of water consumed worldwide goes into supporting agriculture.

Central Asia provides a classic example. Two rivers here feed into the Aral Sea: the Syr Darya and the Amu Darya. Water has been withdrawn from these rivers for 50 years to support a system of cotton monoculture. This form of agriculture in turn supported a system of social and political control.[10] It kept people on the farms, and more importantly allowed governments to control the economy and, essentially, to control people's livelihoods. It ensured that they would not have the chance to challenge the government (especially during the Soviet era) because they had certain basic needs provided for.

This cotton monoculture desiccated the Aral Sea, basically dividing it into two lakes—a northern lake and southern lake. Today there are still more demands being made on the water system. The conflict in Afghanistan has ended, and steps have been taken to rebuild its feeble economy. Afghanistan contains about 17 percent of the water resources in the Aral Sea Basin. There has been a lot of interest among the international community in helping Afghanistan redevelop the upper watershed for agriculture as the basis for economic reconstruction there. But this will have tremendous effects on the downstream states such as Uzbekistan and Turkmenistan which are highly dependant on this water for agriculture.

More importantly, the rivers in Central Asia are fed by glaciers in the mountains of Kyrgyzstan and in Tajikistan. Climate change could accelerate snowmelt, which would affect the seasonal flow. This would affect thousands of livelihoods downstream, as it could have an impact upon the current cropping patterns.

In short, these systems are all interlinked. If we just focus on the snowmelt and what is happening with the glaciers, we will fail to see what is happening downstream in Uzbekistan and Turkmenistan. Here you find a system of cotton farming, which in turn is linked to government and social control. If these countries find themselves forced to restructure their entire economies so as to be able to deal with climate change, they may find themselves facing increased social unrest.

To conclude, when we think about water and conflict, and how these two relate, it is important that we think about the complex ways natural processes and social processes are linked. Thank you.

ENDNOTES - Weinthal

1. E.g., see Peter H.Gleick, "Water and Conflict: Fresh Water Resources and International Security," *International Security,* Vol. 18, 1993, pp. 79-112.

2. UNDP, *Human Development Report 2006—Beyond Scarcity: Power, Poverty and the Global Water Crisis,* available from *hdr.undp.org/hdr2006.*

3. Erika Weinthal *et al.,* "The Water Crisis in the Gaza Strip: Prospects for Remediation," *Ground Water,* Vol. 43, No. 5, 2005, pp. 653-660.

4. Pritt J. Vesiland, "Water: the Middle East's Critical Resource," *National Geographic,* Vol. 183, No. 5, 1993, pp. 38-70.

5. Note, for example, the case of Rwanda. After the genocide, a peace treaty was drawn up. It did not, however, deal with disputes over land, precisely because these were so acrimonious.

6. UNDP; Human Development Report 200.

7. E.g., see Weinthal, pp. 653-660.

8. Charles Vörösmarty *et al.,* "Global Water Resources: Vulnerability from Climate Change and Population Growth," *Science,* Vol. 289, No. 5477, 2000, pp. 284-288.

9. In recent literature, a lot of attention has been focused on battle deaths. In fact, there has been a decline in battle deaths since 1946. Hazem A Ghobarah, Paul Huth, and Bruce Russett, "Civil Wars Kill and Maim People, Long after the Fighting Stops," *American Political Science Review,* Vol. 97, No. 2, 2003, pp. 189-202.

10. Erika Weinthal, *State Making and Environmental Cooperation: Linking Domestic and International Politics in Central Asia,* Cambridge, MA: MIT Press, 2002.

On Climate Change and Infectious Disease: Implications for Political Destabilization and Conflict

Andrew Price-Smith
Colorado College

On Etiology and Emergence.

In the 21st century, novel pathogens are currently "emerging" at the rate of approximately one new agent per annum. Emerging diseases often are the result of "emergent properties" wherein antecedent variables (e.g., population density, speed of transport) combine in unusual and unforeseen ways that facilitate the emergence of a given pathogen which then becomes endogenized within the human ecology. The classic modern example of such emergent properties leading to viral proliferation is the severe acute respiratory syndrome (SARS) coronavirus which appeared in Guangzhou, China, in late 2002, and subsequently spread throughout the Pacific Rim nations. In that particular case, this virulent coronavirus spread from its natural reservoir in east Asian bat populations, into palm civets. The variant of the virus that infected civets was transmissible among humans, amplified by elements of the human ecology such as the "wet markets" of East Asia, the closed environments of modern hospitals which amplified degrees of infection, and modern jet airplane technology that facilitated the rapid spread of the virus throughout the Pacific theatre. Individually these disparate variables would not predict the emergence of epidemic disease; however, when combined together, the SARS contagion of 2002-03 resulted.

The dynamics of contagion frequently exhibit such emergent properties,[1] and the relations between pathogen, human host, and vectors of transmission (e.g., mosquitoes) are central to both the transmissibility and lethality of any given manifestation of contagion. Furthermore, epidemics and pandemics exhibit nonlinearities and threshold dynamics. For example, pathogens may simmer in a given population for some time, but once the rate of transmission passes from <1 to >1, the proliferation of the pathogen may then increase on an exponential scale. Diseases also exhibit high levels of interactivity, and the capacity for co-infection. The classic example is HIV which destroys the host's immune system, and thereby facilitates colonization by other pathogens (e.g., tuberculosis) that ultimately kill the host.

What, then, is the relationship between climate change, infectious disease, prosperity, and political stability and security? The complexity of such interactions is enormous, and so we begin with the relations between climate and disease, focusing on malaria in particular.

Data provided by the IPCC regarding changes in precipitation from 1900-2000 indicate enormous variance on a global scale. Certain regions, such as the arctic and sub-arctic regions of the northern hemisphere, the northeastern sector of south Asia, and Eastern Australia are clearly enjoying increased levels of precipitation. Certain vectors of disease, (such as mosquitoes and snails) thrive in wet environments. Consequently, increases in precipitation will induce the proliferation of vectors, and thereby increase the transmission rates of certain pathogens such as malaria and schistosomiasis.

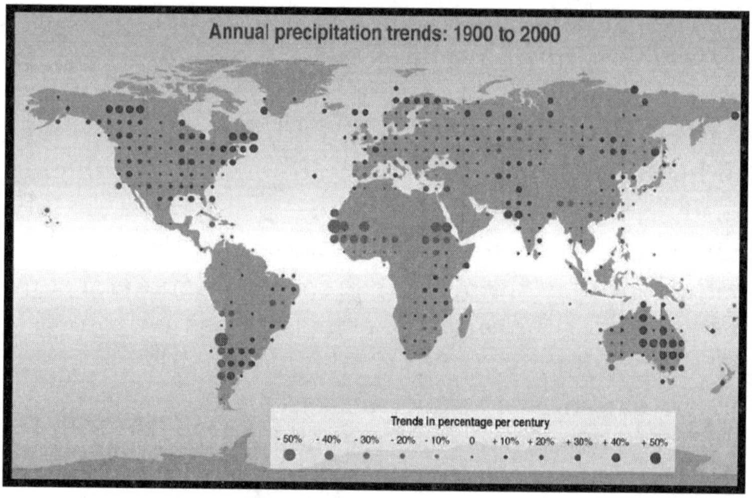

Source: Intergovernmental Panel on Climate Change (IPCC, 2006, available at *www.ipcc.ch/*, accessed on April 2007.

Figure 1.

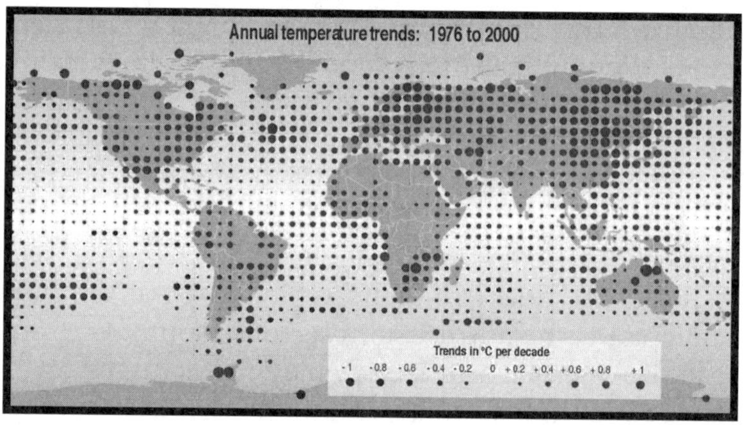

Source: Intergovernmental Panel on Climate Change (IPCC), 2006, available at *www.ipcc.ch/*, accessed on April 2007.

Figure 2.

Pathogens and their vectors of transmission are often highly sensitive to changes in temperature as well. IPCC data from 1976-2000 clearly indicate increasing temperatures for much of the surface of the planet, with the greatest increases evident in the temperate to polar regions. As isotherms shift toward the polar regions, this will expand the latitudinal range of the vectors in question (i.e., anopheles mosquitoes) and thereby permit the expansion of malaria in previously nonmalarious zones. Similarly, increasing surface temperatures permit the movement of malaria in higher altitudes than before. For example, Nairobi has historically been nonmalarial due to its altitude, but in recent years increases in temperature have seen the pathogen moving into the region. The temperature-induced expansion of malaria is problematic because it exposes novel populations, who often lack any genetic or acquired immunity to the pathogen. Thus, the mortality and morbidity in such regions may be much higher than in zones where malaria is endemic.

Increasing temperatures also affect the biting rate of vectors. As temperatures rise, the vectors (mosquitoes) feed with greater frequency, and therefore increase the transmission rate of the *plasmodium* (the parasite) into human populations. Furthermore, increasing temperatures also affect the extrinsic incubation rate of the pathogen, such that it replicates within the gut of the vector at a greatly augmented rate. Thus, under conditions of higher temperatures, there are greater numbers of plasmodium within the vector, and the vector bites with much greater frequency.[2] On a macro level, all of this means that as temperatures increase, the burden of disease (e.g., malaria) is likely to increase to a significant degree. Precipitation and Sea Surface Temperatures (SSTs) are strong predictors of malarial incidence.[3]

In the case of cholera, increasing SSTs are highly correlated with the growth of algal blooms. The blooms move across oceans courtesy of dominant currents and winds and function as vectors of transmission of the *vibrio*. Thus, we see a long-term empirical association between SSTs and the incidence of cholera. In the case of cholera, we have also seen that incidence is responsive to the modulation of the *El Nino* Southern Oscillation (ENSO), with preliminary evidence from case studies carried out in Bangladesh.[4] There is also considerable evidence of thresholds and nonlinearities, such that warming temperatures may produce minor and linear increases in *vibrio* incidence until a threshold point is reached, after which the numbers of the pathogen increase at an exponential scale.

Schistosomiasis is a frequently lethal disease induced by parasitic blood flukes, and it is prevalent in tropical and temperate zones. The vector of the parasite is the snail (*oncomelania*) which thrives under conditions of increased precipitation, and within the temperature range of 15.3 degrees Celsius (C) to an optimal temperature of 30 degrees C. The balance of available evidence suggests that global climate change (GCC) will shift the distribution of the vectors into new regions, and thereby afflict previously uninfected populations. A caveat however, the IPCC data clearly indicate that certain regions (e.g., West Africa) are becoming increasingly arid, which is inimical to the vector. Consequently, those zones that witness declining precipitation levels will see a decline in the incidence of schistosomiasis in their respective populations. In those regions that exhibit increasing precipitation coupled with increasing temperature, we are likely to witness augmented geographic zones of transmission and increased frequency of transmission within those regions. Thus, GCC will result in winners and losers, dependent upon the particular pathogen in question, and its sensitivity to aridity and temperature.[5]

Economic Outcomes.

The economic historian Robert Fogel won the Nobel Prize in economics in 1994 for his analysis of the hypothesis that population health was the central driver of economic productivity.[6] If health promotes prosperity, then disease erodes productivity and wealth. At the microeconomic level, disease erodes productivity through mechanisms such as the debilitation of workers, increased absenteeism, increased medical costs, reduced savings and investment, and the premature death of breadwinners. At the sectoral level, disease imposes a particular burden upon those sectors of the economy that are labor-intensive, such as agriculture and resource-extraction, and thereby imposes a relatively greater effect upon the economies of the developing world.

The impact of malaria is illustrative at the macroeconomic level. Sachs and Malaney estimate that for those countries where malaria is endemic, the pathogen generates a 1.3 percent drag on their gross domestic product (GDP) growth rate, per capita/per annum. Further, Gallup and Sachs estimated that a 10 percent decline in malaria incidence resulted in a 0.3 percent increase in the growth rate of GDP per capita/per annum. McCarthy estimated that malaria imposed a drag on the GDP growth rate of affected nations, at the level of 0.25 to 0.55 percent per annum.[7] In case studies of individual nations, malaria control has resulted in greater prosperity for the polity in question. For example, malaria control measures in Zambia resulted in a $7.1 billion increase to that nation's economy.[8]

The burden of infectious disease falls primarily upon the poor and middle classes, and therefore as the burden of disease increases in certain regions, it will likely exacerbate both the perceived and real level of economic inequities between socio-economic strata. Historically, such perceptions of inequity have led to periods of social and political destabilization.[9] On a global scale, GCC-induced increases in the burden of disease will exert a drag on the global economy, and the perpetuation of poverty within the less developed countries.

Assessments of the economic burden of a given illness (e.g., malaria) are complicated by the lack of adequate surveillance infrastructure throughout much of the developing world where the disease is endemic.[10] Moreover, the complexity of measuring the economic impact of GCC-induced infectious diseases is augmented by the interactivity of various pathogens in a given population. For example, the population of country X may be increasingly beset by increased incidence of malaria, dengue fever, and schistosomiasis, and certain individuals may exhibit co-infection with one or more pathogens.

Pathogens may also erode the functionality and efficacy of the state as well. For example, disease-induced economic stagnation (or contraction) of the macro economy will consequently reduce tax-based revenues available to the state. Diminished revenues will in turn impede the state's capacity to provide public goods and services (e.g., education and law enforcement) to its population. This may in turn reduce the populace's perceptions of the legitimacy of the state. In the domain of human capital, disease may further erode state capacity by debilitating and/or killing trained and skilled personnel, thereby reducing institutional resilience and efficacy.[11]

On Poverty, Instability and Conflict.

The association between poverty, political destabilization, and outright conflict is complex. In particular, there is an endogeneity issue regarding the direction of causality. However, we can make some preliminary observations at this point. First, various iterations of the State Failure Task Force conducted empirical investigations and determined that infant mortality (as a measure) is a strong empirical predictor of state failure.[12] Ted Gurr argued that increasing levels of poverty induced a psychological state of deprivation (perceived injustice) that often led to intrastate conflict.[13] This hypothesis that conditions of deprivation (both real and perceived) led to civil strife was supported by Deininger, and low levels of the Human Development Index are associated with conflict in Indonesia.[14] Other political scientists have found that poverty combines with ethnic fragmentation to produce intrastate conflict.[15] Charles Tilly has argued that inequities are directly associated with intrastate conflict.[16] Further, there is empirical evidence that social polarization leads to conflict, and that conflict may function as a "coping strategy" for those populations confronted with extreme levels of economic deprivation.[17] Convincing arguments take the form of the state weakness hypothesis wherein deprivation combines with a weakened state to offer both the motive and the opportunity for political violence, with evidence from numerous case studies.[18] Political scientists have also hypothesized that increased levels of infectious disease may lead to conflict between sovereign states.[19] Although there is evidence that contagion leads to political acrimony and trade disputes between nations, there is no evidence that infectious disease results in war between nations.[20] Despite the proliferation of literature to support the hypothesis that economic deprivation generates political violence at the intrastate level, additional cross-national empirical analysis, using time-series data, is required. That said, the balance of existing evidence supports the hypothesis.

Conclusions.

Pathogens function as *stressors* that impose burdens on both populations (i.e., society), and upon the structures of the state itself. Historical analysis of the stresses generated by epidemic disease demonstrate that pathogens have exacerbated pre-existing conflicts between socio-economic classes, between ethnicities, and between those of different religious affiliations; and have frequently induced conflicts between states and societies.[21] Thus, the GCC-induced proliferation of disease may facilitate socio-political destabilization, particularly in the weak states and impoverished populations of the developing world. However, such destabilization is contingent upon several factors, it is pathogen-specific, and it depends upon existing socio-economic and political cleavages within the polity in question. Areas at risk of such disease-induced destabilization include the subtropical to temperate zones as tropical pathogens and their attendant vectors expand into these contiguous zones to affect immunologically naïve populations. Thus, we should be concerned about nations in South Asia, Central and East Asia, Southern Africa, and South America.

The effects upon the security of the United States will be *indirect*. However, in the post September 11, 2001 (9-11) era, we now recognize that weak and failed states in the developing world may generate externalities (such as terrorism) that threaten the material interests of the dominant powers of the international system, including the United States.

In conclusion, further research is required to flesh out the complex chain of possible causation that I have detailed above. This will require the formation of interdisciplinary teams of both social and natural scientists who will then model the impacts of climate change upon disease, and the consequent effects upon the economic and political domains. This might involve the compilation of a time-series dataset across a representative sample of countries. One obvious problem involves modeling the long-term processes of climate change. However, we might use the ENSO effect to model how short-term changes in climate induce variance in disease incidence, and then observe the resulting economic and political impacts over the very short term.

ENDNOTES - Price-Smith

1. For an in-depth discussion, see Andrew Price-Smith, *Contagion and Chaos, Disease and Political Discord in the Era of Globalization*, Cambridge, MA: MIT Press, 2008, forthcoming.

2. See Paul Reiter,"Climate Change and Mosquito-Borne Disease," *Environmental Health Perspectives*, Vol. 109, 2001, pp. 141-161; R. S. Kovats, *et al.*, "Early Effects of Climate Change: Do They Include Changes in Vector-Borne Disease?" *Philosophical Transactions: Biological Sciences*, Vol. 356, No. 1411, 2001, pp. 1057-1068; P. R. Hunter, "Climate Change and Waterborne and Vector-borne Disease," *Journal of Applied Microbiology*, Vol. 94, No. 1, 2003, pp. 37-46; M. Van Lieshout *et al.*, "Climate Change and malaria: Analysis of the SRES climate and socio-economic scenarios," *Global Environmental Change*, Vol. 14, 2004, pp. 87-99; J. Patz, "A human disease indicator for the effects of recent global climate change," *Proceedings of the National Academy of Sciences*, Vol. 99, No. 20, 2002, pp. 12506-12508; and Tony McMichael, *Human Frontiers*,

Environments and Disease, New York: Cambridge University Press, 2001.

3. M. C. Thompson *et al.*, "Use of Rainfall and Sea Surface Temperature Monitoring for Malaria Early Warning in Botswana," *American Journal of Tropical Medicine and Hygiene*, Vol. 73, No. 1, 2005, pp. 214-221.

4. X. Rodo *et al.*, "ENSO and cholera: A nonstationary link related to climate change?" *Proceedings of the National Academy of Sciences*, Vol. 99, No. 20, pp. 12901-12906; and J. Patz, "Human Disease Indicator."

5. Zhou *et al.*, "Potential Impact of Global Warming on the Transmission of Schistosomiasis," *Chinese Journal of Epidemiology*, 2002; G. J. Yang *et al.*, "A potential impact of climate change and water resource development on the transmission of schistosoma japonicum in China," *Parassitologia*, Vol. 47, No. 1, March 2005, pp. 127-34; P. Steinmann *et al.*, "Schistosomiasis and water resource development: Systematic review, meta-analysis, and estimates of people at risk," *Lancet Infectious Diseases* Vol. 6, No. 7, 2006, pp. 411-425; G. J. Yang *et al.*, "Effect of temperature on the development of Schistosomiasis japonicum within Oncomelania hupensis, and hibernation of O. hupensis," *Parasitology Research*, Vol. 100, No. 4, March 2007, pp. 695-700.

6 . Robert Fogel, *The Conquest of High Mortality and Hunger in Europe and America: Timing and Mechanisms*, NBER Historical Working Paper 16, September 1990.

7. D. McCarthy *et al.*, *The Growth Costs of Malaria*, NBER paper 7541, February 2000.

8. J. Utzinger *et al.*, "The economic payoffs of integrated malaria control in the Zambian copperbelt between 1930 and 1950," *Tropical Medicine and International Health*, Vol. 7, No. 8, pp. 657-677, August 2002.

9. Price-Smith, *Contagion and Chaos*.

10. E. Worral *et al.*, "The burden of malaria epidemics and cost-effectiveness of interventions in epidemic situations," *American*

Journal of Tropical Medicine and Hygeine, Vol. 71 (2 suppl.) 2004, pp. 136-140; and E. Worral *et al.*, "Is malaria a disease of poverty? A review of the literature," *Tropical Medicine and International Health,* Vol. 10, No. 10, October 2005, pp. 1047-1059.

11. An expanded analysis of the pernicious effects of disease on the state can be found in Andrew Price-Smith, *The Health of Nations*, Cambridge, MA: MIT Press, 2002.

12. D.C. Esty *et al., State Failure Task Force Report I and II,* Washington DC: Science Applications International Corporation (SAIC), 1995 and 1998.

13. Ted Gurr, *Why Men Rebel,* Princeton, NJ: Princeton University Press, 1970.

14. Klaus Deininger, "Causes and Consequences of Civil Strife: Micro-Level Evidence from Uganda," *Oxford Economic Papers,* Vol. 55, 2003, pp. 579-606; H. Malapit *et al., Does violent conflict make chronic poverty more likely? The Mindanao experience,* Working paper, 2003, available from *chronicpoverty.org/pdfs/conferencepapers/malapit.pdf.*

15. W. Easterly and R. Levine, "Africa's Growth Tragedy: Policies and Ethnic Divisions," *Quarterly Journal of Economics*, Vol. 112, No. 4, 1997, pp. 1203-1250; B. Korf, "Functions of violence revisited," *Journal of Conflict Resolution,* Vol. 49, No. 4, August 2005, pp. 451-482; Steven Wilkinson, *Votes and Violence: Electoral Competition and Ethnic Riots in India,* New York: Cambridge University Press, 2004.

16. Charles Tilly, *Durable Inequalities*, Berkeley, CA: University of California Press, 1998. See also Frances Stewart, "Root causes of violent conflict in developing countries," *British Medical Journal,* Vol. 324, February 9, 2000, pp. 342-345; A. Langer, *Horizontal Inequities and Violent Conflict: The Case of Cote d'Ivoire,* Oxford, UK: Centre for Research on Inequality, Human Security, and Ethnicity, University of Oxford, 2004, available from *crise.ox.ac.uk/pubs/workingpaper13.pdf*; L. Mancini, *Horizontal Inequality and Communal Violence: Evidence from Indonesian Districts,* Oxford, UK: Centre for Research on Inequality, Human Security, and Ethnicity, University of Oxford, 2005, available from *crise.ox.ac.uk/pubs/workingpaper22.pdf.*

17. J. Esteban and D. Ray, "Conflict and Distribution," *Journal of Economic Theory*, Vol. 87, 1999, pp. 379-415; C. Boix, "Political Violence," Paper prepared for the Yale Conference on Order, Conflict, and Violence, Yale University, April 30, 2004; M. Humphreys and J. Wienstein, "Handling and Mishandling Civilians in Civil War," *American Political Science Review*, August 2006; and P. Verwimp, "An economic profile of peasant perpetrators of genocide: Micro-level evidence from Rwanda," *Journal of Development Economics*, Vol. 77, No. 2, August 2005, pp. 297-323.

18. Colin Kahl, *States, Scarcity, and Civil Strife in the Developing World*, Princeton, NJ: Princeton University Press, 2006; and Thomas Homer-Dixon, *Environment, Scarcity and Violence*, Princeton, NJ: Princeton University Press, 1999.

19. P. W. Singer, "AIDS and International Security," *Survival*, Vol. 44, No. 1, 2002, pp. 145-158; and Price-Smith, *Chaos and Contagion*.

20. Price-Smith, *Chaos and Contagion*.

21. See Peter Baldwin, *Contagion and the State in Europe, 1830-1930*, New York: Cambridge University Press, 2005; Alfred Crosby, *Ecological Imperialism: The Biological Expansion of Europe, 900-1900*, Cambridge, UK: Cambridge University Press, 1986; Richard Evans, *Death in Hamburg: Society and Politics in the Cholera Years*, New York: Penguin, 2005; J. N. Hays, *The Burdens of Disease: Epidemics and Human Response in Western History*, New Brunswick, NJ: Rutgers University Press, 2003; William McNeill, *Plagues and Peoples*, New York: Doubleday, 1976; Friedrich Prinzing, *Epidemics Resulting from Wars*, Oxford, UK: Clarendon Press, 1916; Terence Ranger and Paul Slack, eds. *Epidemics and Ideas: Essays on the Historical Perception of Pestilence*, Cambridge, UK: Cambridge University Press, 1992; Charles Rosenberg, *The Cholera Years*, Chicago: University of Chicago Press, 1987; and Sheldon Watts, *Epidemics and History: Disease, Power and Imperialism*, New Haven, CT: Yale University Press, 1997.

Climate Change, Population Movements, and Conflict

Timothy J. McKeown
University of North Carolina, Chapel Hill

Climate change will directly affect the operating environment of military forces through such specific changes as the melting of polar ice packs and the thawing of permafrost. However, its impact on international security will arguably be greater through its pervasive and complex effects on the globe's human societies, especially on population movements. As the 2005 example of Hurricane Katrina illustrates, extreme weather events sometimes trigger large, unplanned population movements. Even when the effects are felt in the form of less dramatic but lengthier departures from established patterns, sizeable emigrations can be triggered when local conditions reach a point where local economic and social support systems begin to break down. The experience of eastern Oklahoma during the Dust Bowl period of the 1930s is a good example of this.

In this section,[1] I summarize the array of effects on societies identified in current climate change research based on the 2007 report of Working Group II of the Intergovernmental Panel on Climate Change (IPCC).[2] Next, I discuss effects on the United States, then the results of a project that has developed a procedure for representing each nation's vulnerability to climate change by a climate change vulnerability index. The index numbers are a simple way to depict how climate change might confer relative advantages or disadvantages on a number of powers. A fourth section summarizes research on how population movements —a likely response to extreme weather or to climate changes rendering some regions less habitable—are related to the occurrence of violent political conflict. I conclude with a discussion of the implications for national security if the forecast changes in climate and associated population changes do indeed take place. It is, of course, possible that some forecasts are unduly pessimistic, others unduly optimistic, and that some significant effects simply have not been anticipated at all. However, all national security planning takes place in an environment of uncertainty, and enough research has been conducted that many of the forecasts have a substantial basis in historical data and experimental observations.

Current Assessments of the General Impacts of Climate Change.

Even among highly educated members of the U.S. population, the most common understanding of climate change is to view it as a long, gradual process producing gradual changes, with the whole system readily stoppable or reversible once we believe that we are certain about our understanding of climate change dynamics.[3] That is a faulty understanding for several reasons, but the one that is most salient from a national security standpoint is that human vulnerability to climate change is highly likely to make itself felt not though the cumulative effects of long, gradual, smooth processes, but by sudden, extreme events. The recently completed Report of Working Group II of the Fourth Assessment of the IPCC finds that extreme weather conditions will generally become more common over the 21st century. This is significant because "Climate change vulnerabilities of industry, settlement and society are mainly related to extreme weather events rather than to gradual climate change."[4] Unfortunately for predictive purposes,

> Vulnerabilities to climate change depend considerably on specific geographic, sectoral and social contexts. They are not reliably estimated by large-scale (aggregate) modeling and estimation. . . . The significance of climate change (positive or negative) lies in its interactions with other non-climate sources of change and stress, and its impacts should be considered in such a multi-cause context.[5]

Thus, highly specific and detailed forecasts are subject to high levels of uncertainty. Although they are sometimes useful in sensitizing an audience to possibilities that they had not previously considered, the fact that most such forecasts are inevitably inaccurate can also lead an audience to conclude that forecasters are "crying wolf" simply in an effort to attract attention. If treated more modestly, as a source for a series of "rule of thumb" claims, extant research on climate change does provide helpful insights on how human societies are likely to be affected. The vividness of highly specific and detailed predictions is lost, but a certain level of credibility is gained.

The kinds of claims that current research supports are summarized below. Taken from the 2007 findings of Working Group II, this array of forecast social impacts illustrates a wide variety of climate change impacts beyond increases in average temperatures. It also suggests that who is affected is a product of economic factors—who has access to what resources—as well as political ones—which governments are going to adopt and successfully implement policies designed to mitigate the impacts of climate change? The issue of who is affected is also not just a matter of geography, but also of demographics, with some groups—especially those dependent on the smooth functioning of the larger social system—especially vulnerable to negative events.

Climate Phenomena	Other Casual Factors	Projected Impact	Zones,Groups Affected
a) Increased frequency of extreme events			
Tropical cyclones, storm surge	Population density, land use in flood-prone areas; flood defenses; institutional capacities	Increased vulnerability in storm-prone coastal areas; possible effects on settlements, health, tourism, economic and transportaion systems, buildings and infrastructures	Coastal areas, settlements and activities; regions and populations with limited capacities or resources; fixed infrastructure; insurance sector
Extreme rainfall, riverine floods	As above, plus drainage infrastructure	As above, plus drainage infrastructure	As above, plus flood plains
Heat or cold waves	Building design and internal temperature control; social contexts; institutional capacities	Increased vulnerabilities in some regions and populations; health effects; change in energy requirements	Mid-latitude areas; elderly, very young, ill or very poor populations
Drought	Water systems; competing water uses; energy demand; water demand constraints	Water resource challenges in affected areas; shifts in locations of population and economic activities; additional investments in water supply	Semi-arid and arid regions; poor areas and populations; areas with human-induced water scarcity
b) Changes in mean levels of climate variables			
Temperature	Demographic and economic changes; land-use changes; technological innovations; air pollution; institutional capacities	Shifts in energy demand; worsening of air quality; impacts on settlements' livelihoods depending on melt water; threats to built environment from thawing permafrost soils in some regions	Very diverse, but more vulnerability in places with populations with more limited capacities and resources for adaption
Precipitation	Competition from other regions or sectors. Water resouce allocation	Vulnerabilities in some areas to effects of precipitation increases (e.g., flooding, but could be positive) and in some areas to decreases (see drought above)	Poor regions and populations
Saline intrusion	Trends in groundwater withdrawl	Increased vulnerabilities in coastal areas	Low-lying coastal areas, especially those with limited capacities and resources
Sea-level rise	Trends in coastal development, settlement, and land uses	Long-term increases in vulnerabilities of low-lying coastal areas	As for saline intrusion
c) Abrupt climate change in general			
	Demographic, economic, and technological changes; institutional developments	Possible significant effects on most places and populations in the world, at least for a limited time	Most zones and groups
Source: Adapted from IPCC (2007: Table 7.3)			

Table 1. Projected Social Impacts of Climate Change.

These effects are not easily summarized. They show a wide variety of undesirable changes, an equally wide variety of affected areas and groups, as well as substantial uncertainty about the timing, location, and magnitude of specific events. Because climate change as a threat is multifaceted, broad-scale, and nonspecific, we can only be certain that we will be responding to it, without necessarily knowing which of the above processes will be most important at any given time or location.

Effects on the United States.

The 2007 working group II assessment suggests that the most significant effects on U.S. residents will likely be:

- effects on coastal regions due to rising oceans, rising ocean temperatures, more variability in weather, and more severe weather. Increased variability of weather not only means more flooding, but also more droughts.
- effects on river basins due to greater variability in precipitation, declines in precipitation in arid and semi-arid areas, and greater winter and smaller summer flows from snow packs and glaciers.
- effects on forest, food, and fiber agriculture as a result of changes in average weather conditions, and the occurrence of more extreme weather conditions. In middle to high latitudes, crop yields might modestly increase, and adaptation to climate change might minimize negative effects, but this is less likely to be successful in lower latitudes and in arid or semi-arid regions, and for larger increases in temperatures. "Over

the 21st century, pressure for species to shift north and to higher elevations will fundamentally rearrange North American ecosystems."[6]

- effects on public health as high temperatures and extreme weather likely lead to more heat-related mortality, pollution, storm-related fatalities and injuries, and infectious diseases. The geographic range of various pests and diseases will shift, with tropical and sub-tropical varieties moving to higher latitudes.

Overall, the forecast effects are significant and mostly negative. However, that is true for most countries. How is the United States likely to fare compared to other powers?

Cross-National Comparisons.

Although climate change has and will continue to impose costs around the globe, these costs are not borne equally. National security analysts rightly focus on factors affecting not merely absolute but also relative capabilities. To the extent that nations differ in their exposure to adverse effects, their capacity to adapt to the changes or to mitigate the damages that the changes inflict, their relative advantages or disadvantages in military or nonmilitary competition, will be affected.

Maplecroft is a British firm that specializes in research, management consulting, training, and other forms of organizational development that bear upon corporate social responsibility at the global level. The firm has developed an index of climate change vulnerability that provides a useful starting point for considering the question of how differences in national vulnerabilities to climate change translate into the conferring of security advantages or disadvantages on the world's nations. Their index relies on a small number of general measures widely available for the nations of the world. Scores on the three dimensions of coastal vulnerability, inland vulnerability, and health-related vulnerability are averaged to generate a single summary statistic that serves as a guide to relative vulnerability. (Details on the construction of the index are provided in the Appendix). The skeptical reader might note that it is but one of many such possible approaches to constructing an index, and that at this point there is no basis for strongly preferring it to other plausible contenders. The skeptical reader would, of course, be correct. However, the necessity to begin the consideration of the questions involved, its ready availability, and transparent documentation all commend it as a useful starting point.

Social Processes Triggered by Climate Change.

The human impacts of these changes that have the most implications for social disruption—and hence for national security—fall into two related categories. The first are the generic consequences of extreme weather conditions triggering large-scale natural disasters: large-scale loss of life; of habitation; of essential services such as drinking water, sewage treatment, and energy; and of the capacity of the governments in question to provide order and emergency assistance, and begin the process of recovery. The second category is all of the impacts that are a result of relatively large and sudden migrations of human populations. These might be the result of a large natural disaster, but can also occur as a result of gradual deterioration leading to the crossing of some threshold, after which the situation takes a dramatic turn for the worse.

Country Exposure	Overall Impact	RANK* (among 189 countries)	Coastal Exposure	Inland Exposure	Health
Poland	Low	12th	Medium	Medium	Medium
Japan	Medium	35th	High	Medium	Low
Australia	Medium	44th	High	Medium	Low
Russia	Medium	46th	Medium	Medium	Low
China	Medium	73rd	High	Low	Medium
Germany	High	93rd	Medium	Extreme	Low
Brazil	High	95th	Medium	Medium	High
India	High	103rd	High	Low	Extreme
Iran	High	155th	Medium	High	High
Canada	Extreme	168th	High	Extreme	Low
USA	Extreme	174th	High	Extreme	Low
Iraq	Extreme	185th	Medium	High	Extreme

* From least to most vulnerable. See Appendix for detailed explanation of the rankings.

Source: Based on data provided by Maplecroft to accompany their map of climate change impacts. The map, as well as maps and data covering greenhouse gas emissions, carbon resources, renewable energy use, and energy security risk, can be found at *http://www.global-risks.com/content/maps/.*

Table 2. Selected Countries Ranked by Forecast Vulnerability to Effects of Climate Change.

The possibility that climate change could trigger large population movements, and that these movements would spawn political conflict and perhaps even large-scale violence, has already begun to attract attention within the U.S. national security community. In a 2003 report, consultants to the Defense Department's Office of Net Assessment explored the national security implications of a relatively abrupt climate shift triggered by sudden changes in ocean currents.

In such a situation, large scale population movements could be triggered by food shortages due to decreasing global agricultural production, decreased fresh water supplies, and disruptions in transoceanic movements of oil and coal.[7]

Responding to natural disasters is part of the repertoire of organizational skills that governments in developed countries supposedly possess. While in that sense climate change does not pose a novel challenge, the frequency and the magnitude of such disasters are likely to increase, and such developments might impose demands on the existing infrastructure of civilian and military governmental organizations, as well as nongovernmental organizations, that these organizations are not presently equipped to meet.

Similarly, while migration has been a constant feature of human history, the proportion of the world's population migrating at any given time might well substantially increase, and patterns of migration might well depart from those experienced in the recent past. The report of Working Group II is cautious in assessing these possibilities, merely noting that:

> large numbers of displaced people are a likely consequence of extreme events. Their numbers could increase, and so could the likelihood of their migration becoming permanent, if such events increase in frequency. Yet, disaggregating the causes of migration is highly problematic, not least [because] individual migrants may have multiple motivations and be displaced by multiple factors.[8] . . . Estimates of the number of people who may become environmental migrants are, at best, guesswork since (a) migrations in areas impacted by climate change are not one-way and permanent, but multi-directional and often temporary or episodic; (b) the reasons for migration are often multiple and complex, and do not relate straightforwardly to climate variability and change;

(c) in many cases migration is a longstanding response to *seasonal* variability in environmental conditions, it also represents a strategy to *accumulate* wealth or to seek a route out of poverty, a strategy with benefits for both the receiving and original country or region; (d) there are few reliable censuses or surveys in many key parts of the world on which to base such estimates (e.g., Africa); and (e) there is a lack of agreement on what an environmental migrant is anyway.[9] . . . [R]ising ethnic conflicts can be linked to competition over natural resources that are increasingly scarce as a result of climate change, but many other intervening and contributing causes . . . need to be taken into account. For example, major environmentally-influenced conflicts in Africa have more to do with relative abundance of resources, e.g., oil, diamonds, cobalt, and gold, than with scarcity.[10] This suggests caution in the prediction of such conflicts as a result of climate change.[11]

U.S. history presents us with a small-scale analogue to possible global migratory patterns in the form of the drought-induced exodus of farm families from Oklahoma in the 1930s. The case is instructive partly because later droughts in eastern Oklahoma that were comparable in severity to the "dustbowl" drought did not spur nearly the same level of emigration.[12] Changes in the economic vulnerability of the population to drought, the attractiveness of easy-to-reach destinations, and the capacity of local institutions to counteract the effects of the drought or adjust to them all help to account for differences in human responses to highly similar climatic events. Similarly, Sen and Davis[13] have shown that famines can be as much the consequence of social, political and economic inequalities that affect access to food, as they are of pestilence or extreme weather.

The Political Consequences of Demographic Changes.

Nazli Choucri pioneered the modern study of the political impact of migration and other demographic changes. Her summary of the importance of demographic factors in violent conflicts[14] is still a useful summary of the factors that are politically significant. (See Table 3 below.)

Although all of these factors play some role in some situations, the most frequently important one is segmental divisions, a category that captures ethnic or religious differences. Events since 1984 (especially since the break-up of the Soviet empire) have, if anything, further confirmed the significance of segmental divisions as sources of violent political conflict. However, Table 3 is also a useful reminder that such segmental divisions hardly exhaust the sources of violent conflict.

Migrating populations often have distinct characteristics that differentiate them from either the population from which they are departing or the ones that will be their new neighbors. Probably the single most important one from a military security standpoint is that those most likely to migrate are young males[15]—the primary recruitment pool for militaries and guerrilla armies the world over. Conversely, a migrating population with large numbers of children or the elderly means that such a group will require more aid, or a relatively quick and successful integration of employable family members into local communities to provide them with the earnings that they will require to feed and house their dependents. It also means that relatively fewer will be of fighting age and joining militaries or paramilitaries or gangs, and that those who are of the correct age might feel the tug of family

Variable	Total Number of Cases	Background Factor	Minor Irritant	Major Irritant	Central Importance	Sole Determinant
Population size*	22	3	15	1	2	1
Absolute size	16	8	4	2	2	0
Size in relation to resouces	19	1	8	7	2	1
Population change	19	8	6	5	0	0
Absolute rate of growth	25	10	10	3	2	0
Differential rates of growth	9	3	5	1	0	0
Population Distribution	35	7	10	13	5	0
Rural/urban distribution	16	3	10	3	0	0
Population density	7	2	3	2	0	0
Spatial location in relation to resources	18	3	4	8	3	0
Spatial location in relation to borders	15	4	5	4	2	0
Population Movement	30	11	6	8	5	0
Population Composition	32	1	6	11	13	1
Sex distribution	4	2	2	0	0	0
Age distribution	10	3	3	4	0	0
Segmental divisions	31	2	1	10	12	6
Level of knowledge and skills	26	4	9	10	3	0

* General factors are in bold type, specific factors are not. The number of cases where a specific factor affected outcomes is typically less than the number of cases where the general factor did, because for some cases a given specific factor played no role.

Table 3. Importance of Demographic Factors in Violent Conflict.

obligations. Assessing the political impacts of any given flow of people across borders will require going well beyond simply counting heads and figuring out how many thousand calories of food will be needed each day by each group.

Migration becomes easier to the extent that a migrating population already has ties to their destination region. Early migrants play an important role in welcoming later arrivals, providing them with familiar products and services, help in adapting to their new locale, and an immigrant community into which they can integrate. Thus, migration patterns tend to exhibit some persistence or inertia.[16] Knowing historical migration patterns has some value for forecasting where subsequent migrating populations will go. However, widespread alteration of the earth's ecosystems might also disrupt currently observed migration patterns and create new ones.

Conclusions.

The possibility that climate change will lead to more instances of extreme weather not easily forecast far in advance, and that such weather will bring about sudden and large movements of affected populations, is probably the most important security challenge that climate change presents in the near term. Over a long period, more gradual and longer-term processes are forecast to drive various ecosystems across thresholds into new configurations where previous patterns of human settlement and economic activity might no longer be possible. Taken together, these forecasts suggest a substantially greater demand for disaster management by the array of government agencies, international organizations, and nongovernmental organizations that now engage in such activities. The expected increase in population movements, especially sudden and unplanned movements, implies the likelihood of an elevated level of "normal" social and political turmoil, and a correspondingly higher level of global political turbulence.

The disruptive effects of climate change will likewise have broad effects on national capabilities. In an absolute sense, most nations will experience significant effects; in a relative sense, the resulting distribution of gains and losses will tend to disadvantage some more than others. The Maplecroft index suggests that North America will suffer greater net negative impacts than most other regions,[17] suggesting that there is a strong case to be made for U.S. leadership in efforts to minimize the negative effects of greenhouse gas emissions simply in terms of traditional national security concerns about relative national capabilities.

While coping with the implications of high levels of greenhouse gases will require a global response beyond the capabilities of any single national government, the United States over much of the post-1945 era has several times confronted situations where national objectives could only be met by international action. The United States has often responded to these situations by becoming a leader in fashioning international institutions to create international public good in the realms of military security and economic prosperity. Both U.S. and global interests would be well-served by a similarly energetic and imaginative effort to create the international institutions that will be necessary in order to cope with the looming consequences of climate change.

Appendix

The Maplecroft Climate Change Vulnerability Index

The Maplecroft Climate Change Index is intended to capture nations' relative levels of exposure to the physical and health impacts of climate change. The index has three components: coastal exposure, inland exposure, and health exposure. The Index is simply the unweighted mean of each country's three scores.

Coastal Exposure.

This component quantifies the exposure of each country to rises in sea levels and increasing coastal flooding during storms. It is based on the following indicators:

- Percentage of land below five meters above sea level.[18] Weight: 40%.
- Percentage of population living below five meters above sea level.[19] Weight: 40%.
- Total number of people affected by tropical storms between 1975 and 2005 as a percentage of the population.[20] Weight 20%.

Inland Exposure.

This component quantifies the exposure of each country to extreme temperature events, inland flooding, and food availability. It is based on the relative risk factors presented in the World Health Organization (WHO) publication, "Comparative Quantification of Health Risks."[21] The risk factor is defined as future changes in global climate attributable to increasing atmospheric concentrations of greenhouse gases.

Projections of the extent and geographical distribution of climate change within each nation were calculated by WHO by applying various emissions scenarios to the Hadley Centre's HadCM2 global climate model.[22] The HadCM2 model generates projections of changes in temperature and other climatic factors which have been verified by back-casting. Average climate conditions during 1961-90 provide the baseline observations.[23]

The data used for calculating the inland exposure component came from the high estimate for 2030 using the IPCC IS92a scenario (unmitigated greenhouse gas emissions). The scores were calculated by summing the relative risk factors for extreme temperature events, inland flooding, and food availability.

Health Exposure.

The health exposure component was calculated using data for estimated mortality attributable to climate change from McMichael *et al.*[24] WHO calculated mortality attributable to climate change for five specific health outcomes selected on the basis of their observed sensitivity to temporal and geographic climate variation, their importance in terms of mortality and burden of disease, and the availability of quantitative global models.[25] The five health outcomes and the class that they proxy are as follows:

Outcome class	Outcome
Direct effects of heat and cold	Cardiovascular disease deaths
Food-borne and water-borne diseases	Diarrhea episodes
Vector-borne diseases	Malaria cases
Natural disasters	Unintentional deaths and injuries
Risk of malnutrition	Nonavailability of recommended calorie intake

Mortality estimates were based on observations of the effects of recent shorter-term climate variation (e.g., the effects of daily or interannual climate variability on specific health outcomes) or the present (e.g., climate as a determinant of current disease distribution), or on specific processes that may influence health states (e.g., parasite and vector population dynamics in the laboratory, determining the transmission of infectious diseases). These quantitative relationships were then applied to future climate scenarios.

Adjustments were made for possible changes in vulnerability through biological or socioeconomic adaptation. Estimates of future effects were then interpolated back to give an approximate measure of the effect of climate change that has occurred since 1990 on mortality and burden of disease in the year 2000. This gives a measure of the magnitude and distribution of health impacts of climate change. The health exposure component was calculated by indexing the calculated deaths per million population for the combination of the above five health outcomes.

ENDNOTES - McKeown

1. This paper was written specifically for this edited volume. At the conference, the author of this article summarized the presentation sent by the scheduled speaker who was unable to attend.

2. Intergovernmental Panel on Climate Change, Fourth Assessment Report, Working Group II, "Impacts, adaptation, and vulnerability," 2007, available from *ipcc-wg2.org*. Hereafter IPPC, 2007.

3. J. D. Sterman and L. B. Sweeney, "Understanding public complacency about climate change: Adults' mental models of climate change violate conservation of matter," *Climatic Change*, Vol. 80, pp. 213–238.

4. IPCC, 2007, Summary of Chapter 6. The report also notes that, "Gradual changes can also be associated with thresholds beyond which impacts become significant."

5. *Ibid.*, Summary of Chapter 7.

6. *Ibid.*, p. 619.

7. P. Schwartz and D. Randall, "An abrupt climate change scenario and its implications for United States national security," Prepared for the Department of Defense, Office of Net Assessment, October 2003.

8. R. Black, "Environmental refugees: Myth or reality?," New Issues in Refugee Research, Working Paper 34, United Nations High Commissioner for Refugees, Geneva, 2001.

9. J. Unruh, M. Krol, and N. Kliot, "Environmental Change and Its Implications for Population Migration," *Advances in Global Climate Research,* Vol. 20, 2004; H. Eakin, "Institutional change, climate risk, and rural vulnerability: Cases from Central Mexico," *World Development,* Vol. 33, 2006, pp. 1923-1938.

10. J. Fairhead, "Achieving sustainability in Africa," in R. Black and H. White, eds., *Targeting Development: Critical Perspectives on the Millennium Goals,* London, England: Routledge, 2004, pp. 292-306.

11. Working Group II 2007, Box 7.2.

12. R. A. Warrick, "Drought in the Great Plains: A case Study of Research on Climate and Society in the USA," in J. Ausubel and A. K. Biswas, eds., *Climatic Constraints and Human Activities,* Oxford, England: Pergammon Press, 1980, pp. 93–124.

13. A. Sen, *Development as Freedom,* Oxford, England: Oxford University Press,1999; and M. Davis, *Late Victorian Holocausts: El Niño Famines and the Making of the Third World,* London, England: Verso, 2001.

14. Nazli Choucri, *Population Dynamics and International Violence: Propositions, Insights, and Evidence,* Lexington, MA:

Lexington Books, 1974, p. 102, cited in N. Choucri, "Perspectives on population and conflict," in Choucri, ed., *Multidisciplinary Perspectives on Population and Conflict,* Syracuse, NY: Syracuse University Press,1984, p. 16.

15. R. McLeman and B. Smit, "Migration as an adaptation to climate change," *Climatic Change,* Vol. 76, 2006, pp. 31–53.

16. D. S. Massey and R. M. Zenteno, "The dynamics of mass migration," *Proceedings of the National Academy of Sciences of the United States of America,* Vol. 96, No. 9, 1999, pp. 5328-5335.

17. One other region where a cluster of "extreme impact" countries can be found is East Africa as far south as Mozambique, the Arabian Peninsula, and South Asia from Iraq to Pakistan.

18. Centre for International Earth Science Information Network (CIESIN), *National Aggregates of Geospatial Data: Population, Landscape and Climate Estimates,* New York: Columbia University, 2002, available from *sedac.ciesin.columbia.edu.*

19. *Ibid.*

20. Collaborating Centre for Research on the Epidemiology of Disasters (CRED), The OFDA/CRED International Disaster Database, Data Version v05.09, Brussels, Belgium: Universite Catholique de Louvain, 2005, available from *www.em-dat.net.*

21. A. J. McMichael *et al., Climate change and human health: final report to the Department of Environment, Transport and the Regions,* London, England/Maastricht: London School of Hygiene and Tropical Medicine/ICIS, 2000.

22. S.F.B. Tett, T.C. Johns, and J.F.B. Mitchell, "Global and regional variability in a coupled AOGCM," *Climate Dynamics,* Vol. 13, 1997, pp. 303–323.

23. M. New, M. Hulme, and P. D. Jones, "Representing twentieth century space-time climate variability. Part 1: Development of a 1961-90 mean monthly terrestrial climatology," *Journal of Climate,* Vol. 12, 1999, pp. 829-856.

24. A. J. McMichael *et al.*, "Global climate change," *Comparative Quantification of Health Risks*, Geneva, Switzerland: World Health Organization, 2004.

25. McMichael *et al.*, 2000.

Commentator

David Gilmartin
North Carolina State University

I really appreciate being asked to do this. I have to say global climate change is not the area that I am most familiar with. I am a historian, and have worked in particular on the politics and the history of water in India and Pakistan, but from that angle I do have some ideas on some of the interesting presentations we just heard.

To begin with, I would like to go back to the first panel this morning, and make a couple of comments as a way of framing how we might think about these papers and their relevance to the overall topic. The papers this morning on climate change were, for me, fantastic, because they were a real introduction to a lot of what is going on in the scientific community on this; and particularly to the science.

One thing that really struck me, though, was how significant the "global" part of the equation was. We saw the image this morning of the globe taken from the moon several decades ago. This global dimension is, I think, one of the reasons why global climate change is such an important and critical issue in popular politics. The power of this issue lies in the way that it creates a framework for thinking about humanity as a single community. It puts the role of scientists, and an international community of scientists, at the very center of the discussion.

And it is a quite powerful force this way, particularly as juxtaposed to the notion of globalization as an economic issue. Economics, of course, also defines world-wide processes, though it does so through theories that posit notions of integration through people pursuing their distinctive self-interests. Climate change has, of course, been significantly influenced by the changes that economic globalization has produced, particularly in the wake of the industrial revolution. Understanding environmental change thus undoubtedly depends on understanding the operation of the world economy. But the importance of studying climate change in long-term perspective lies precisely in the fact that it also allows us to stand outside these economic processes and look at the human community in different sorts of terms. Like the image of the globe taken from the moon, it creates an image of human community that transcends the realities of the world economic system, and thus transcends also economic perspectives.

As we try to figure out how to approach the topic of climate change, we thus need to think about how to put these perspectives together. We need to understand what is happening in the world economy to understand climate change and to understand possibilities for action in dealing with it. But the significance of environmental perspectives lies in imagining a human community in relationship to the environment that stands outside such strictly economic perspectives, and outside the conflicts over resources that strictly economic valuations generate. If these perspectives are not put together, no projection or models can fully capture the past and point us toward the future. One way to do this is through history.

Now let me turn to the three papers, all of which I enjoyed. I would like to put them a bit further into historical context: Without knowing how we got to where we are, it is pointless to try to figure out what might happen in the future.

Let me start with Erika Weinthal's presentation, which I found very challenging and enjoyable. We talk about environmental changes caused by climate change. But, of course, as her paper illustrates, there have been vast environmental changes associated with water that are linked to what some would call the "Irrigation Revolution" of the last 100 or 150 years, which in some ways, utterly constrain the ways that we can respond to climate change now. In fact, the very meaning of water scarcity, as she points out, is conditioned by the ways that water has been used for irrigation in the creation of irrigated societies. If people were to step back and imagine a world outside history, many people would say the massive development of extensive irrigation on arid lands in the last century makes no sense whatsoever. But, historically vast expenditures and investment in irrigation have transformed huge areas of the world and created the water environment that exists in the world today. This is therefore an environment that today we live with.

What I really liked about Erika's paper was her point that now—and if one looks at the history, one can see this as well—irrigation development has been deeply embedded in particular forms of power, in particular kinds of social structures, and in notions of state authority and how it operates in irrigated societies. All these things now create a huge constraint, and create a framework in which conflict may occur. Large-scale irrigation systems, almost by their very nature, make water scarce. But we need to remember the history of how it got that way. When we think about what water scarcity means, we should bear in mind that scarcity is not solely, or even primarily, a concept defined by nature, but one defined by the histories of the ways that states have sought to extend their power over land. As Erika makes clear, trying to understand water scarcity, as a technical problem, without understanding history, is a fool's errand. Water scarcity may have potential implications for international conflict, but these implications cannot be addressed apart from the processes of state-building that have produced them.

It would also be useful, I think, to put Andrew Price-Smith's presentation—which I also learned a lot from—into historical context. To take a huge historical frame for this, one could say that the greatest change in human history, which opened the doors to the worst ravages of disease on human societies, was the Agricultural Revolution. This was one of the critical moments, of course, in the development of human societies. At the same time, it was a moment which made possible, because of the environmental changes associated with it and particularly, as some have argued, the relationship between domesticated livestock and human beings, a vast expansion of the place of endemic diseases in human life. This reminds us that disease has a very complex relationship to what we would see as some of the most important advances in human history, and so, as we think about the future, one would have to ask questions about that. No easy generalizations about disease and human progress are possible.

I think that Price-Smith's point about poverty, disease, and conflict is a very interesting one. Let me suggest, however, that, although disease undoubtedly has an impact on state capacity and poverty (and thus for potentially engendering conflict), from my own perspective, the causation in the other direction has historically been far more powerful. Poverty and its attendant malnutrition has historically been one of the great enablers of the spread of disease. So, I find it a bit problematic to talk about the impact of disease on state capacity without emphasizing, more basically, how poverty itself is related in myriad ways to state structures and to the relationship of states in the international economy. As far as poverty and international conflict goes, one need only look at the history of the world economy to see the range of arguments on this. Some of the most powerful historical arguments, going back to John Hobson's critique of imperialism at the beginning of the 20th century, argued that there was a direct connection between inequality in European societies (and poverty) and what one might call Europe's military adventurism overseas. One could well argue that that relationship, though it is a controversial one among historians, has by no means gone away. In fact, the relationship between domestic inequality and overseas militarism and conflict remains an important issue in the discussion about America's overseas military policy even today.

The final presentation was also very illuminating for me. Here again, though, I think a historical framework would really help. I am not sure how one can gain perspectives on climate change, demography, and conflict without putting the subject into the broader context provided by the history of world population change and its relationship to industrialization and the world economy. The theory of the demographic transition would provide an invaluable perspective. Demographic change has, of course, been intimately related to the vast economic changes both before and after the industrial revolution which, after all, have been critical to processes of global warming.

I would just stop there with those few comments and thank the panel givers very much for some very interesting papers.

Discussion

Q: Migration from the northeastern sectors of Brazil into Sao Paolo and Rio de Janeiro arguably led to a coup by the Brazilian army. Does this kind of case study help us understand how demographic shifts caused by climate change might become national security threats?

Weinthal: I cannot really speak to Brazil, but I can point to other areas where natural processes disrupted livelihoods. These are not necessarily related to shifts in climate. In Rwanda, for example, it was misuse of land that played a role. It is often hard to prove that climate change is a direct cause of conflict. Whether or not an environmental event becomes politicized and has a disruptive outcome depends on the broader social and economic context.

Q: A lot of questions surround global precipitation cycles. It seems to me that societies are growing more sensitive to short-term precipitation deficits. Is that true? Does the security risk come from pressures on existing water supplies caused by growing populations or from long-term climate change and what it will do to water supplies?

Weinthal: A bit of both. Short-term provision of water is a concern. It is a problem, though, if we limit ourselves to short-term solutions. Jordan, for example, has limited sources of water as it is. If the region is going to become more arid, it is essential to engage in long-term planning. Relying on fossil groundwater will do in the short, but not the long-term.

Q: Do we need to just build more dams to capture rain when it does fall and use this during the dry periods? Or must we engage in social engineering?

Weinthal: Again, a bit of both. There are ways to augment water supplies. One way is desalination (a technique proposed for Gaza). But this is problematic—desalination is energy-intensive. It might work for Saudi-Arabia, but for Gaza? And, of course, this is only going to contribute to the problem of climate change which is adversely affected by the use of fossil fuels. Another way is to use recycled waste water in agriculture or introduce drip irrigation. But here, too, there are problems, as experience has shown. In Central Asia, Israelis introduced drip irrigation in the cotton fields to help improve the situation in the Aral Sea. They laid down the lines at night, and the next morning when they got to the fields, everything was ripped up. Here was a solution that was economically appropriate and environmentally beneficial. But it meant doing things differently, and the locals found it disruptive. If you modify irrigation patterns in a society which is based on irrigation, you will break down social networks. That adds a whole level of uncertainty and scares people.

Q: How have migration patterns changed with changing conditions? We need to know this so that we can bounce them against some of these global climate change trends we are beginning to see.

McKeown: We do have an overall sense of what has driven the rise and fall of the movement of populations over the last 2 centuries. That is largely a story about economic development, the opening of the new world, and declining transportation costs. The story becomes more complicated as we get into the early 20th century, when states started to regulate the international movements of people. What we do not have is a broader picture of how political change, particularly political violence, creates longer, larger trends. We tend to study the relationship between political violence and the movement of people on a case by case basis, and our conclusions about it tend to be heavily context-specific, so in that regard, I am afraid I don't have a lot to offer you. There are people now, however, who speculate that we are entering a world where state capacities are going to tend to decline across the board. This will likely lead to more unauthorized and quasi-illegal movement of people across borders.

CHAPTER 3

U.S. NATIONAL DEFENSE

This chapter looks at the security threat from a more narrowly U.S. perspective. Dr. Joshua Busby, Dr. John T. Ackerman, and Dr. Kent Butts discuss the direct and indirect repercussions of global climate change: how growing inequalities might impact the struggle against extremism, what financial and other demands these might place on the United States, and how U.S. strategic relationships with the world at large might change. The authors consider both threats and opportunities.

Climate Change: Complicating the Struggle against Extremist Ideology

Kent Hughes Butts
U.S. Army War College

I would like to share a few thoughts with you today on environmental security and climate change and how they relate to combating terrorism. At the end of my talk, I will suggest how we might bring all the elements of national power to bear on the problem. There are a number of themes we should consider as we proceed:

1. Stability. Terrorism is rooted in regional instability which has posed the major threat to U.S. national security interests since the end of the Cold War.

2. Opportunities. Change, even global climate change, brings opportunities.

3. Legitimacy. Being able to meet the needs of the people is essential for any government.

4. Confidence building measures. Climate change phenomena provide opportunities to establish communication and cooperation both multilaterally and intrastate, and deescalate tensions.

5. The elements of power. Secretary of State Madeleine Albright was quoted as saying, "What's the point of having this superb military . . . if we can't use it?" We can now see the downside of using the military element of power to effect regime change without also having a sound plan for using the other elements of national power to create a stable political system capable of meeting the demands placed on it. The United States must find creative ways to use all the elements of national power to deal with problems resulting from global climate change if it is to successfully manage terrorist insurgencies and ensure regional stability.

Environmental Security and the National Security Strategies.

Climate change is best understood through the lens of **environmental security**. Not all environmental issues are environmental security issues. Dr. Alan Hecht of the Environmental Protection Agency tells us that "Environmental Security is a process whereby solutions to environmental problems contribute to national security objectives." A flood in which there is no human involvement or loss of resource wealth is not a security issue. However, a flood that displaces 100,000 people across a national border, overwhelming the resources of a fragile neighboring state, and destabilizing the region is a security issue. Global climate change is a subset of environmental security and because it can be a force multiplier for existing tensions, it will complicate the ability of the United States to manage threats to regional stability and preserve U.S. national security interests.

Climate change brings opportunity because today's science is so good that it can predict where the effects of climate change will be most pronounced and threaten regional stability. As the awareness of climate change effects becomes more widespread within the security community, the opportunity to synchronize the programs of the U.S. interagency in addressing them will offer an advantage.

The National Security Strategy (NSS) report was designed to preclude U.S. involvement in conflicts around the world where its interests had not been properly defined. The 1986 Goldwater-Nichols Department of Defense (DoD) Reorganization Act mandated this annual report to Congress. The NSS should clearly define U.S. national security interests and a strategy for achieving the objectives necessary to preserve those interests. The best and most useful NSS reports have defined U.S. national security interests by region, clarifying any threats to those interests, articulating a strategy for attaining objectives, and the resources necessary to do so. This is not a Defense document; it addresses the use of all elements of national power, including the interagency community, which largely represents the available resources.

Each of the agencies, then, is supposed to develop its own strategic documents delineating how it will support the NSS intent. DoD, for example, develops the National Military Strategy and the National Defense Strategy, which provide guidance for DoD policymakers and guidance for military forces. A similar process takes place within the Department of

State (DoS) providing guidance to the Embassies within which reside the important Regional Environment Offices or Hubs.

The NSS will typically define U.S. national security interests to include combating terrorism, democratization, development of market economies, regional stability, the nonproliferation of weapons of mass destruction (WMDs), and a strong economy. Most of these have been on the list for the last 15 years, though their prominence may vary. Two of them will figure prominently in today's talk—the fight against terrorism and the effort to create regional stability. Environmental security and climate change, although very important to these two interests, are not a priority of the current administration.

Environmental security has figured prominently in most of the national security strategies generated over the last 20 years. First consider this quote from President Reagan's 1988 document: "The dangerous depletion or contamination of natural endowments of some nations—soil, forest, water, and air—will create potential threats to the peace and prosperity that are in our national interests as well as the interests of the affected nations." This NSS emphasizes the relevance of the land, man/land relationship to national security.

In the Bush and Clinton administrations' National Security Strategies of the 1990s, environmental security was recognized as an important variable: "The stress from environmental challenges is already contributing to political conflict" (1991); and "[Environmental security issues] pose a direct or indirect threat to U. S. National Security Interests" (1997); and "Environmental security issues often trigger and exacerbate conflict" (1997). Note the language used in these documents. As several of the speakers today have already noted, controversy does exist as to causality; if a statistical analysis cannot prove causality, the test will say so. Critics will often seize upon the results to suggest that the issue is not relevant to security, and thus does not warrant policymaker attention. This is particularly true in regions of existing tensions. Because policymakers are interested in relevance, not overstating causal relationships is important. Well-respected academic literature may rule out water as a cause of violent conflict, while a policymaker like Ariel Sharon will be quoted as saying that the 1967 War actually began 2 years earlier when Israel attacked the Arab Jordan River Headwater Diversion scheme to prevent the diversion of the Jordan River headwaters. Thus, it is often more useful and avoids needless controversy to characterize environmental challenges, to include climate change events, as triggers, stressors, multiplier effects, or having exacerbated pre-existing tensions.

This issue has a direct bearing on the issue of terrorism and climate change. Does poverty cause terrorism? In Washington, DC, it is common to hear the phrase, "Poverty doesn't cause terrorism." This is justified in part because there are areas of the world where there is poverty and no terrorism. However, in Southeast Asia where terrorism and insurgency have a long history, national leaders such as Philippine President Gloria Arroyo and Malaysian Prime Minister Abdullah Ahmad Badawi speak with great conviction about the need to address the link between poverty and terrorism. Overreaching led in part to the current National Security Strategy stating that "Poverty does not, in and of itself, cause terrorism." This diminishes the perceived U.S. Agency for International Development (USAID) role as a resource for addressing terrorism. However, stating that poverty erodes governmental legitimacy and contributes to conditions that terrorists seek to exploit invites the use of USAID's substantial capacity to address the terrorist insurgency.

Finally, note the emphasis in the current National Security Strategy, "We will harness the tools of economic assistance, development, aid, trade, and good governance to help ensure that new democracies are not burdened with economic stagnation or economic corruption" (2006). Once again, we see recognition at the highest level that development, economics, empowerment, and building capacity for our allies are important; and it is the relationship of climate change to these variables that defines its greatest relevance to regional security.

The Military in Mitigation and Adaptation.

It is not surprising to find that the military is often the largest agency in a government, with a correspondingly dominant budget. Many developing countries have bloated militaries that are remnants of the Cold War or past military threats to the state. Apropos, downsizing of these militaries is politically difficult, and many downsizing initiatives have resulted in unintended regime change. Thus, the regime may continue, but at a substantial opportunity cost for other agencies struggling to meet the demands placed upon the regime. In regions where climate change is predicted to affect weather patterns, crop production, disease, and clean water availability, and may do so disproportionately between states and within states, involving the military in providing support to other agencies tasked with addressing these issues may mean the difference between maintaining governmental legitimacy and a failed state.

In a diagram of Maslow's hierarchy of needs for a developing state, stewardship of the environment would likely be placed at the pinnacle of the triangle, as a goal to fulfill only after other demands upon the political system have been met. In fact, it really should be at the base, because sustaining the environment underpins success in the other seemingly more important base categories such as food and shelter, security from violence, economic growth, and development. Meeting all of these demands will contribute to maintaining a government's legitimacy and providing for the human security needs of its people. Failure to do so leaves the government vulnerable to alternative visions and, in some countries very important to U.S. national security, to extremist ideology and terrorist insurgency. The military can contribute to developing a country's resilience to climate change effects. Unlike many government agencies, the military has a presence on the frontier and relatively good transportation assets and fuel; it also has technical, medical, and civil works capabilities and a training and education program. Developing the capacity of the military to help in climate change mitigation and adaptation efforts in support of other agencies could enhance resilience, reduce systemic stress, and gain legitimacy for the government and the military.

In order to have a stable world and continued prosperity, it will be necessary to find creative solutions for the effects of global climate change. In seeking these solutions, the host nations' militaries should be seriously considered.

Addressing global climate change is not just a matter for DoD. DoS is also very much involved in issues of environmental security. Secretary of State Warren Christopher, at the request of then Vice

President Gore, created the system of environmental hubs. These bring the resources of powerful regional bureaus within DoS to bear on solving environmental security issues. There are 12 of these world-wide. In our work at the Center for Strategic Leadership, we work closely with the Combatant Commands to help them develop environmental security programs. When we put together a regional conference or activity on environmental security for those Commands, our first action is to contact the relevant environmental hub and seek their guidance and support. This has led to many areas of interagency cooperation and, quite often, to the Commands following the lead of the State environmental hubs in their efforts to develop regional capacity and promote multilateral cooperation. Currently we are working with the DoS environmental hub in Brasilia in response to a Brazilian Army request to support an environmental security conference for South America that focuses on alternative fuels. When we do this, we are bringing the substantial capabilities of the military element of power to bear on climate change issues that are a priority of the U.S. interagency community.

Climate Change and Combating Extremist Ideologies.

Climate change is affecting the efforts of the United States to combat the global insurgency and its underlying extremist ideologies in two ways: First, it provides a strategic communication windfall for the insurgency, allowing extremists and critics of the United States to claim that the United States does not care about the welfare of other countries. These countries, they say, must struggle with the rising energy costs and global warming that directly result from the high U.S. per capita consumption of energy resources. Second, climate change is complicating the ability of countries to meet the needs of their people, thus enhancing the appeal of extremist ideology by creating underlying conditions terrorists may exploit.

The role of climate change in security ranges from the global to regional to local levels. The Russian geopolitical quest for dominance of the energy fields of the Arctic was spawned by the unexpectedly rapid melting of the polar ice cap. The coastal resources of Florida are threatened by saltwater incursion because of the drought affecting the Southeast; and there are many traditional security concerns that directly affect U.S. national security interests that are being influenced by climate change. These concerns are highlighted by two important studies. One was published in 2007: *National Security and the Threat of Climate Change* by the Center for Naval Analyses Corporation. The leader of the study's research team, Tom Morehouse, discusses the highly influential study in Chapter 6. The National Intelligence Council (NIC) is preparing a National Intelligence Estimate on this subject. According to the NIC, the variables include: Conflict; Failed States; Terrorist Opportunities; Economics; Energy; Social Unrest and Migration; Humanitarian Crisis; and Unrest in Islamic Countries.

We heard excellent presentations this morning about the cause of climate change. However, these two studies are not being conducted to contest critics who maintain climate change is not anthropogenic. They are moving beyond the causal debate and instead saying, "We represent the security community and we are convinced that we are facing an imminent threat, and we need to be prepared to deal with this threat."

Many of the variables that these studies address are going to affect the ability of the United States to manage the terrorist insurgency and compete with extremist ideologies overseas.

In 2003 when visiting the U.S. Army War College, Secretary of Defense Donald Rumsfeld conceded that the United States was not doing as well as it could in the war of ideas.

> If I were grading, I would say we probably deserve a "D" or a "D-plus" as a country as to how well we're doing in the battle of ideas that's taking place in the world today. I'm not going to suggest that it's easy, but we have not found the formula as a country for countering the extremists' message. The strategy must do a great deal more to reduce the lure of the extremist ideology by standing with those moderate Muslims advocating peaceful change, freedom, and tolerance.

His point is salient. To defeat terrorist ideology, the United States requires a good message and the support of those who are trying to prove it to be accurate and appealing. Climate change undermines both of those objectives. To understand how, it is useful to review the U.S. strategy for combating terrorism.

In February 2003, the National Security Council published the *National Strategy for Combating Terrorism* (NSCT). This strategy had four pillars: defeat terrorist organizations with global reach; defend the interests, citizens, and territory of the United States both domestically and overseas; deny terrorist groups sponsorship, support, and sanctuary; and diminish the underlying conditions terrorist seek to exploit. The latter two are directly related to climate change. The United States sought to encourage other states to meet the terrorist threat by building their capacity and will, while winning the war of ideas by promoting state and regional stability through political, social, and economic development.

The United States found it easy to take the moral high ground against the oppressive Soviet enemy of the Cold War: its message is no longer well-respected by the developing world. The invasion of Iraq and its mishandled aftermath, with such issues as Abu Ghraib and torture, was a propaganda gift to the insurgents and those opposing the moderate Muslim regimes upon which the U.S. combating terrorism strategy depends. So, too, is the U.S. position on climate change. The United States is perceived by the developing as having double standards double standards. It asks the developing world to preserve its forests to serve as carbon stores, and, though the largest per capita emitter of greenhouse gases, denies its responsibility for creating climate change stresses. These stresses are seen as giving rise to droughts, floods, disease, and the weakening monsoons that no longer feed the mountain glaciers providing essential water to many cities. They also complicate governmental legitimacy by overloading weak political systems and demonstrating the developing government's weak economy and social infrastructure, and its inability to manage the resulting human security problems. The terrorist insurgency is rooted in the developing world and seeks to take advantage of weak or failed states and exploit the resulting ungoverned spaces to train insurgents and exert political control.

The NSCT strategy was replaced by a three pillar strategy developed by Secretary Rumsfeld that seeks to: protect the homeland; disrupt and attack terrorist networks; and counter ideological support for terrorism. The elements of the original strategy are subsumed under these pillars, but the greater emphasis on combating the ideology of extremism places even more importance on strategic communication messages and the critical factor of moderate governments developing the capacity to meet the challenges of climate change.

To achieve success over the insurgency in "the long war," attacks against terrorists and their networks will not be enough; we must think in terms of countering extremist ideologies. While climate change factors, such as the drought related conflicts in Darfur, cause instability, they also create an awareness of the importance of building governmental capacity to address environmental security issues. The new Africa Command embraced the concept that, in order to counter extremist ideology and promote the stability essential to U.S. interests, the military needs to prioritize "enabling" partner nations. If we use all the instruments of power at our disposal to help these partner nations build their own capacities, over time they will be able to undermine ideological support for terrorism. Terrorism is fueled by discontent. Countries where there is drought and instability, where failed states are not meeting the needs of their people, places like Mali and Somalia, are breeding grounds for terrorism. The military recognized that it should actively contribute to the establishment of those conditions that would reduce ideological support for terrorism. Thus, environmental security issues, and the climate change multiplier affects, became priority issues for the military, and building good governance and capacity is a major priority of the new Africa Command.

The importance of governmental legitimacy to managing the terrorist insurgency was perhaps best stated by Lieutenant General Wallace C. Gregson, then

Marine Forces Pacific Commander, in his remarks to the "Underlying Conditions of Terrorism Conference" at the U.S. Army War College in June 2005.

> The center of gravity, the decisive terrain of this war, is the vast majority of people not directly involved, but whose support, willing or coerced, is necessary to insurgent operations around the world. . . . Providing what local governments have not, these insurgents have gained legitimacy, psychologically conditioned these populations, and created an area from which they can safely operate. Winning the hearts and minds of local populations is far more important than killing or capturing people.

As the 9/11 (September 11, 2001) Commission Report states, "When people lose hope, when societies break down, when countries fragment, the breeding grounds for terrorism are created."

Recommendations.

In determining how to craft policy options for mitigating the effects of climatic disruption, one should seek to apply all four elements of national power: economic, diplomatic, informational, and military. How should these resources be applied most effectively and creatively to deal with the impact of global climate change? Rather than responding to crises and possibly being forced to undertake the costly introduction of large combat forces, the United States should be proactive in its approach. In the Clinton administration, the 1997 and 1999 National Security Strategies took a dual approach to promoting U.S. national interests, emphasizing "Shaping the International Environment," as well as "Responding to Threats and Crises." These NSS's specifically detailed the roles of all four of the elements of power in affecting this strategy.

At DoD, Secretary of Defense William Perry had a preventive defense strategy that argued for shaping the security environment and being prepared to respond if necessary. The United States should "promote trust, stability, and democratic reform, and so help to prevent the conditions for conflict and build the conditions for peace." He was supported by Deputy Undersecretary of Defense Sherri Goodman, who testified before Congress saying, "We will mitigate the impacts of adverse environmental actions leading to international instability." Thus, the United States would commit military forces to address soft power activities so that it would not have to fight wars. This proactive approach is essential to dealing with the terrorist insurgency and addressing the underlying conditions that terrorists seek to exploit; conditions that are being worsened by changes in climate.

A valuable approach to this problem would be to apply the concept of resilience that the National Oceanographic and Atmospheric Administration (NOAA) introduced for tsunami preparedness, seeking to actively promote mitigation and adaptation programs aimed at making communities climate change resilient. The U.S. interagency, through the regional USAID Missions, the DoS environmental hubs, and the Combatant Commands; with valuable technical support of the Environmental Protection Agency (EPA), U.S. Geological Survey (USGS), NOAA, and the U.S. Forest Service (USFS); could make meaningful contributions to governmental capacity. Such an approach would build governmental legitimacy, allow communities to contribute to problem solving, and reduce the appeal of extremist ideology and the potential for failed states.

Conclusion.

In summary, we need to think about how existing regional issues are complicated by global climate change, and develop a strong domestic message that makes clear to the American people how destabilizing conditions in other parts of the world threaten U.S. national security interests and our way of life. We need to think about how U.S. foreign policy can be proactive in nature, identifying ways to build the capacity of other countries and regions so that their governments will be able to deal with the problems of instability and climate change themselves, before they lead to destabilizing crises or the election of alternative voices such as Hamas or Hezbollah. If the United States is to preserve its security, it must seize the engagement and strategic communications opportunity presented by the climate change phenomenon. It must build networks of resilient and capable regional states able to manage climate change and other destabilizing variables that would otherwise threaten U.S. national security.

Under What Conditions Could Climate Change Pose a Threat to U.S. National Security?

Joshua Busby
University of Texas, Austin

Introduction.

My interest in national security and climate change started in 2004 when Nigel Purvis and I were at the Brookings Institution. We were commissioned to write a paper on the links between climate and security for the United Nations (UN) High-Level Panel on Threats, Challenges, and Change.[1] In that paper, we looked mostly at the effects of extreme weather events in the developing world and the possibilities for large-scale humanitarian disasters in countries that had little capacity to respond.

When I looked at the environmental security literature, much of it focused on the links between environmental change and violent conflict, and it seemed to me that there was a more expansive set of problems that were relevant to security. At the same time, there was a rather careless use of the security frame to encompass all harms to human welfare. This threatened to make security so amorphous as to be meaningless. Attempts to "securitize" climate change appeared to me to be an effort to figure out how to get U.S. decisionmakers to care about this problem.[2]

I set off to write a new paper on climate change and national security that could stand up to scrutiny from traditional security types, both practitioners and academics. In that paper, I identified a broad set of potential security challenges linked to climate change, and I assessed, in a preliminary fashion, whether these purported links are indeed security challenges for the United States.

You might think that U.S. national security is a parochial subject. However, if U.S. decisionmakers are supposed to care about the problem of climate change and if they think about the world primarily in terms of a national interest lens, then it makes sense, at some level, to look at the problem in the way they see the world. In addition, the United States is the biggest greenhouse gas producer and the strongest military power in the world. Its participation will be necessary if the effects of climate change are to be minimized, and if countries buffeted by climate change are going to get the assistance they need. So, again, I think it makes sense to look at this through a national security lens and from the perspective of U.S. national security in particular.

Climate-Related Security Threats.

To assess whether or not the links between climate and U.S. national security are credible, I sought to develop a concept of security that hews closely to traditional notions of security but could also be open to environmental harm. Before talking about this conceptual framework, let me tell you a bit about the breadth of climate-related potential security problems that I discuss in that paper. They include three classes: The first are direct threats to the homeland, the second are indirect threats to the national interest, and the third are broader security externalities of energy policies.

I consider three kinds of direct threats to the homeland: (1) abrupt climate change, (2) sea level rise, and (3) extreme weather events. I conclude, based on my reading of the scientific evidence to date, that only the third, extreme weather events, currently constitutes a security risk that policymakers are likely to find credible, given their short-term time horizons. We will come back to this later in this talk.

I also consider indirect threats to the national interest that include the extraterritorial effects of climate change on U.S. interests. I explore the connections between (1) climate change and violent conflict in the world; (2) the connections between climate change, natural disasters, and humanitarian tragedies in strategically important areas; and, finally, (3) I consider the links between climate change and U.S. soft power and its international reputation.[3]

I also look at the security externalities of climate change. First, I consider how oil dependence may give rise to two "social bads," contributing to two problems simultaneously, both climate change and the increased risk of terrorism. Second, I also look at how a remedy to climate change, nuclear power, may pose a proliferation risk.

While I have not yet written about this in as much detail, the geo-strategic consequences of reactions to climate change are also of interest. For example, under the Kyoto Protocol, Russia was awarded emissions levels based on the economy it had before the former Soviet Union collapsed. This gave it scope to sell permits it did not need to countries like Canada and Japan that did. As an inducement to ratify Kyoto, Russia was also awarded generous credits for the carbon sequestration potential of its forests. What are we to make of the large transfer of funds that might result? Might they bolster Russia's efforts to tamp down on domestic dissent and throw its weight around in its near abroad? If Kyoto blows apart, perhaps these transfers will never materialize. It would, however, be interesting to consider whether the Japanese and Canadians still intend to purchase credits from the Russians.[4]

As the economic consequences of responses to climate change take on real significance—just think about the European emissions trading program—we can expect to see these issues begin to interface with a broader set of diplomatic challenges, whether it be the U.S.-Indian nuclear deal, technology transfer to India and China, or other issues seemingly far removed such as the fate of the Middle East.

Defining Security and Assessing Risk.

Let me engage in a brief discussion of the conditions under which a new problem constitutes a real *security* risk. Security threats are traditionally associated with armed external attack against a state by another country's military. Security is typically thought of as protecting one's territorial integrity and ensuring survival of the state. But, we also know that security can be more broadly defined. It can include a country's vital interests overseas that, if challenged, the effects could be so severe that the country's way of life would be threatened. For these sorts of overseas interests, such as access to oil, the country would be willing to go to war to defend them.

Although national defense is frequently associated with the military and the use of force, this is not the only way a country can protect its security. Diplomatic and economic means are also part of a country's tools to defend its national interest, as the United States understood quite well when it supported the Marshall Plan after World War II. Increasingly, we realize that attacks from other states are not the only ways a country's security can be threatened. Both terrorism and civil wars remind us that nonstate actors and internal threats may constitute national security challenges.

What about harm that is not the result of intentional human agency—specifically environmental harm? Here we have to imagine threats so severe that a country will find its way of life will be threatened, either in terms of direct threats to the homeland or its extraterritorial interests.

For a direct threat to rise to the level of a security concern, one or more conditions would have to be met. It would have to: (1) threaten the existence of the country, or (2) threaten to decapitate its center of government, or (3) destabilize the government such that its monopoly of the use of force over its territory would be compromised over an extended period of time and over a significant geographic space, or (4) be one that could disrupt or destroy critically important infrastructure, or, finally, (5) lead to such catastrophic loss of life and well-being as to undermine the government's legitimacy.

As I suggested earlier, states also have interests beyond their borders. For an extraterritorial threat to constitute a security risk, it would have to be in a location strategically significant in some way. It would also have to reach thresholds of concern in terms of the stakes involved and/or a potential loss of life. The criteria of "strategically significant places" include: (1) those belonging to allies, (2) those which serve as transportation corridors, (3) those containing sources of raw materials, (4) those where bad consequences could blow-back on the homeland, and others. I also talk about different thresholds that would constitute high stakes and high potential loss of life. Situations likely to have higher stakes are those where there is a possibility of state failure or regional contagion. In terms of the thresholds for high loss of life, I suggest the somewhat arbitrary figure of 10,000 or more civilians at risk of death either from violence or as a result of dislocation from extreme weather events.

There are some places that are not strategically important enough or where the stakes aren't high enough for the problem to constitute a national security risk for the United States. But where the strategic significance is high and the stakes are high, these situations should be considered security threats to the United States.

Let me say a word, though, about dealing with uncertainty. Climate change is a problem with effects some of which are of low or unknown probabilities and/or distant in the future. Policymakers have been resistant to take action because they are fearful of false alarms and over-reaction. Thus, the most politically salient effects of climate change will be those the scientists already think are likely to occur in the coming 2 decades.

Climate Change as a Security Threat.

At this point, let us look at some of the evidence that supports my contention that climate change may constitute a direct threat to the homeland or an extraterritorial threat to our interests.

Direct Threats to the Homeland. In my research into what might or might not constitute a direct threat to the homeland, I looked at abrupt climate change, sea level rise, and extreme weather events.

Abrupt climate change speaks to the threats of the slow-down in the Gulf Stream and other changes that might occur in a matter of a few decades, as we heard this morning from Dr. Corell. Scientists are worried about that possibility, but they don't really have a good handle on how likely these threats are. Their best guess is that they are of low probability or are, at the very least, not likely to happen during this century.[5] So, based on this reading of what the Intergovernmental Panel on Climate Change (IPCC) says about abrupt climate change, I conclude that it is not currently a national security risk for the United States around which policymakers would likely mobilize concern.

I then look at sea level rise. Despite Gore's warning of a 20 foot sea level rise,[6] the IPCC fourth assessment report suggested that, across the range of models, the estimates for predicted sea level rise were between 20 centimeters and 60 centimeters by the end of the century, more or less.[7] As we heard this morning, these estimates don't take into consideration the possible melt and collapse of the Greenland and Antarctic glaciers, which appears to be occurring more rapidly than anticipated. These models also don't take into account the ways in which some areas may have faster and higher sea level rise than others.

But, again, these modest increases in sea level are expected to unfold only over the next 80 years. Thus, coastal populations, particularly in a rich country like the United States, would have plenty of time to adapt, either by moving or preparing countermeasures. So the prospect of New York City or San Francisco under water from rising seas, which would require forced migration and the mobilization of the military to evacuate them, seems rather unlikely, in my view.

Extreme weather events are, in my estimate, a more serious threat to the homeland. Such events include storm surges, hurricanes, droughts, and extremely hot days. While we cannot attribute a single weather event like Hurricane Katrina to climate change, the scientific community now thinks that climate change is likely to make extreme weather events more frequent or (in the case of hurricanes) more severe.[8]

The prospect of more Hurricane Katrinas in the United States could pose a national security risk to our country. At the most basic level, this would be manifest in terms of the diversion of military resources to deal with humanitarian needs and to restore domestic civil order, as we saw in New Orleans. This can take on larger significance if it impairs a country's ability to extend its monopoly of force over its territory for extended periods of time. This is not likely to happen in the United States, but it might happen in weaker, poorer countries, and some are places that we might care about.

At the very least, if the assessment of extreme U.S. vulnerability to climate change is right,[9] then we need to be thinking about the local consequences of rising extreme weather events and what that might mean for disaster preparedness and military mobilization in the United States. We need to be prepared to imagine what would happen if we had as much trouble responding to the disaster as we did in New Orleans. The legitimacy of our government would be questioned. This would not only have local political consequences but would have wider ramifications for the way the United States is perceived on the world stage.

Climate Changes and Overseas Impacts.

In looking at the security consequences of climate change on U.S. overseas interests, I consider the links between climate change and violent conflict, climate change and natural disasters, and climate change and U.S. soft power.

Climate Change and Conflict. Climate change will likely make a number of problems worse, particularly extreme weather events. But does it increase the prospects for violent conflict? As Andrew Price-Smith suggested earlier, increasing deprivation caused by climate change might lead to poverty which in turn would lead to conflict. However, there is a real concern that those causal linkages may not, in fact, be credible. Maybe people just don't fight when they are poor. Maybe they do not have a strong enough sense of grievance to fight. Or maybe they just can't—they are so exhausted by their daily struggle that they just will not take up arms. That said, Mark Levy at Columbia University has recently found a correlation between the variability of rainfall and the enhanced onset of violent conflict.[10] Other studies have found a correlation between refugee flows and conflict as well as natural disasters and conflict.[11] Given that scarce rains, refugee numbers, and natural disasters are all likely to be intensified with climate change, we should be worried and prepared with much more extensive programs to help countries adapt to climate change and respond to emergent crises.

Climate Change and Extreme Weather Events. It is important for us to look at the links between climate change and violent conflict, but arguably more important to look at the links between climate change and extreme weather events. The reason I say this is because far more people are affected every year by extreme weather events than by violent conflict.[12] If we focus solely on armed conflict, we are going to be missing a hugely important part of the problem. In the 2004 piece I wrote with Nigel Purvis, our biggest concern was the prospect of extreme weather events and humanitarian disasters. We anticipated that the

United States would be called upon for humanitarian intervention overseas. Little did we imagine that a year later the United States would be staging such a crisis rescue at home with more than 70,000 soldiers deployed to New Orleans in the aftermath of Hurricane Katrina.[13] I talked about extreme weather events in the context of the United States; these problems are likely to have an even bigger impact on those countries that are less well-prepared, less wealthy, with worse governments than our own. We need to look at where our strategic overseas interests match up with the places most vulnerable to climate change and help those countries prepare.

Climate Change and Soft Power. Finally, we should consider the connections between soft power, U.S. reputation in the world, and climate change. To the extent that our government has not done as much on climate change as other countries in the world would like us to, this has added and exacerbated the problem of our reputation in the world. If you don't think that matters, then it is irrelevant. But if you think what other countries think of us matters, then doing something on issues they care about so that they cooperate on issues we care about could be highly instructional. There are better and worse ways of burnishing our international reputation, and climate change may be an especially costly way to restore our good image in the world. If you think our good image is important, then this is one area where doing less than desired has damaged our interests.

Climate change and national security is a very complex issue deserving of far more detailed discussion and analysis. I hope, however, that this brief survey has given you food for thought.

ENDNOTES - Busby

1. Nigel Purvis and Joshua W. Busby, *The Security Implications of Climate Change for the UN System. Submission to the United Nations High-Level Panel on Threats, Challenges, and Change*, Washington, DC: Woodrow Wilson Center, 2004, available from *www.un.org/secureworld/report2.pdf*.

2. This talk is based on a paper authored by Joshua Busby entitled "Who Cares about the Weather? Climate Change and U.S. National Security." The most recent version, with a much fuller citation list, can be found on *www.utexas.edu/lbj/faculty/joshua-busby*.

3. A new area of concern has emerged since the writing of this 2004 paper, and that is the possibility for national competition over Arctic sea lanes as Arctic Sea ice melts. Elizabeth L. Chalecki of the Fletcher School at Tufts has dealt with this issue. Elizabeth Chalecki, "He Who Would Rule: Climate Change in the Arctic and Its Implications for U.S. National Security," paper presented at the International Studies Association, Chicago, 2007.

4. For an effort to address this dimension, see Gregg Easterbrook, "Global Warming: Who Loses and Who Wins?" *The Atlantic*, 2007.

5. IPCC, *Climate Change 2007: The Physical Science Basis-Summary for Policymakers*, February 2007, available from *www.ipcc.ch/SPM2feb07.pdf*, accessed March 1, 2007.

6. Albert Gore, *An Inconvenient Truth: The Planetary Emergency of Global Warming and What We Can Do About It*, Emmaus, PA: Rodale Press, 2006.

7. IPCC, *Climate Change 2007*. See also Robert J. Nicholls, "Coastal Flooding and Wetland Loss in the 21st Century: Changes under the SRES Climate and Socio-Economic Scenarios," *Global Environmental Change*, Vol. 14, No. 1, 2004.

8. Judith Curry, P. J. Webster, and G. J. Holland, "Mixing Politics and Science in Testing the Hypothesis That Greenhouse Warming Is Causing a Global Increase in Hurricane Intensity,"

Bulletin of the American Meteorological Society, Vol. 87, No. 8, 2006; IPCC, *Climate Change 2007*. An accessible book that looks at the science behind the possible links between climate change and hurricanes is by Chris Mooney, *Storm World: Hurricanes, Politics, and the Battle over Global Warming*, Orlando, FL: Harcourt, 2007.

9. See the Mapcroft map of climate vulnerability at *http://maps.maplecroft.com/*.

10. Marc A. Levy, "Is the Environment a National Security Issue?" *International Security*, Vol. 20, No. 2, 1995. Similar findings were reported by Cullen S. Hendrix and Sarah M. Glaser, "Trends and Triggers: Climate Change and Civil Conflict in Sub-Saharan Africa," paper presented at the Human Security and Climate Change, Oslo, Norway, June 21-23, 2005.

11. On the links between disasters and conflict, see Dawn Brancati, "Natural Disaster Conflicts," paper presented at the American Political Science Association, Philadelphia, PA, 2006. On the links between refugees and conflict, see Idean Salehyan and Kristian Skrede Gleditsch, "Refugee Flows and the Spread of Civil War," *International Organization*, Vol. 60, No. 2, 2000. A somewhat skeptical take on the links between climate, refugees and conflict is Nils Petter Gleditsch, Ragnhild Nordås, and Idean Salehyan, *Climate Change and Conflict: The Migration Link*, New York: International Peace Academy, May 2007, available from *www.ipacademy.org/our-work/coping-with-crisis/working-papers*, accessed July 21, 2007.

12. Between 1990 and 1999, one report estimated that 188 million people per year were affected by natural disasters, six times more than the 31 million affected by armed conflict. UN/ISDR, "Living with Risk: A Global Review of Disaster Reduction Initiatives," Geneva, Switzerland: United Nations Inter-Agency Secretariat of the International Strategy for Disaster Reduction (UN/ISDR), 2003.

13. At its peak, the number of soldiers mobilized in the aftermath of Hurricane Katrina exceeded 70,000, including 22,000 active duty and more than 50,000 of the National Guard (about 10 percent of the total Guard strength). *Hurricane Katrina: Better Plans*

and Exercises Needed to Guide the Military's Response to Catastrophic Natural Disasters, Washington, DC: Government Accountability Office, May 2006, available from *www.gao.gov/new.items/d06643.pdf*, accessed July 7, 2007.

The "Perfect Storm"

John T. Ackerman
USAF Air Command and Staff College

The *Quadrennial Defense Review* (QDR) is a cornerstone of our national defense policy and is created by the Department of Defense (DoD). My plan today is to introduce planning models from within this defense planning document and then use climate change as a case study to test my thesis against the planning models. Essentially, I will take the main points in the QDR and show how they might be applied to create a QDR for climate change.

The QDR 2006 starts with a chapter on "Fighting the Long War." It goes on to discuss how to operationalize the strategy, and then how to reorient our capabilities and forces to fight this long war. By the "Long War," the authors of the QDR mean terrorists. Right now the United States has a strong conventional, symmetrical capability. The DoD wants us to reorient our forces so that we are more capable of fighting asymmetric enemies. Also, the Defense Secretary wants to pay special attention to what he calls an "enterprise activity" —the reorganization of the DoD so that it can become more agile and more cost effective. The QDR suggests that we face four kinds of September 11, 2001 (9/11) post-security challenges—traditional, irregular, disruptive, and catastrophic. It also notes four focal points of interest: defeating terrorist networks, defending the homeland, shaping the choices of countries, and, finally, preventing hostile states and nonstate actors from acquiring or using weapons of mass destruction (WMD). The QDR also draws lessons learned from our experiences in Afghanistan and Iraq. It stresses the need to build partnership capabilities, to take early anticipatory measures, and to engage in unity of effort. And it speaks of the "uncertainty of predictability." These challenges, focal points, and lessons learned can, I think, be applied to the environment we are likely to encounter in the event of global climate change.

Let us start by looking more closely at how the QDR conceptualizes the challenges facing us. It divides the threats into four conceptual challenge areas. **Traditional** threats involve force on force, or threats from conventional or symmetric sources. The asymmetric challenges or **irregular** challenges arise when we become involved in guerilla wars, or counter insurgencies. **Disruptive** challenges occur when the enemy does something that undermines our current capabilities. For example, an advance in biotechnology or directed energy weapons offsets our current advantages. **Catastrophic** challenges takes place when, for example, terrorists or other nonstate actors use WMD or weapons which have WMD-like effects.[1]

Figure 1 is a nice graphic representation that shows the relationships between the four challenge areas and our current capabilities.[2] It makes clear that we have great strengths in one area—meeting traditional challenges. However, we have got to build our capabilities up in other areas. We need to center or balance our capabilities so that we can deal with all four of them equally well.

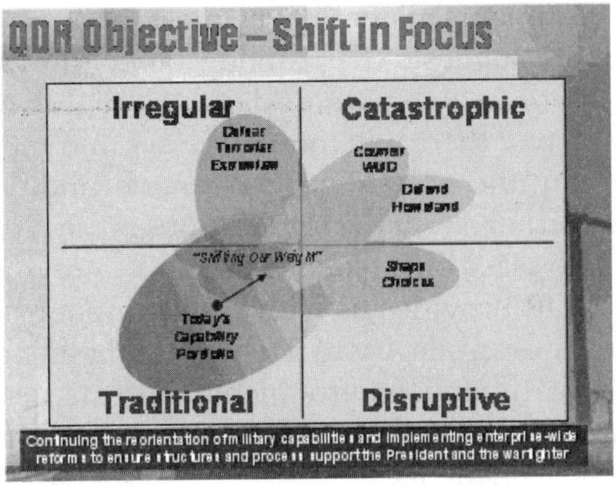

Figure 1.

Applying the QDR.

Can this model help us address climate change? Yes, with modifications. Granted, climate change is a dispersed global threat. By that I mean that it has a wide range of possible consequences, which might include synergistic, nonlinear, and abrupt affects.[3] But I still think that, as the QDR suggests, the thing we want to do is to reorient our defense capabilities. We can apply the challenge area concepts found in the 2006 QDR to good purpose, although we do need to shift from a very specific focus on the war on terror to a broader focus on achieving sustainable security. Sustainable security is a new concept that can be defined as "providing for security in a manner that, at the very least, does not diminish or compromise, and at very best actually enhances an environmentally, socially, and economically sustainable quality of life for future generations worldwide."[4]

Earlier today, you heard about what climate change involves and what causes it.[5] What of the results? The

Intergovernmental Panel on Climate Change (IPCC) 2007 report suggests that, among other things, there will be more frequent warm days, the nights will not be as cool, and there will more hot days and more heat waves.[6] But these are all rather generic, and we need to think further about the implications of such changes if they take place on a global scale.

The QDR provides us with a useful tool or model. We can, for example, try to determine whether a given set of climatic and environmental threats are likely to constitute a traditional, irregular, disruptive, or catastrophic challenge.

So, what would be a **traditional** challenge created by climate change? Well, we're pretty good at addressing heat waves, droughts, and floods[7] here in the United States. We have encountered these challenges before and know how to respond. So these could be construed as falling into this category.

But what happens if we have a lot more of these events that challenge our conventional capabilities from different angles all around the world? Well, we are not really good at enforcing global treaties that might have to be made to deal with climate change. What would be the DoD role in enforcing global climate regimes? Nor are we very good with the concept of ocean acidification,[8] which we just tangentially covered. What happens if our oceans become more acidic, and some of the creatures that live in it become extinct, and the whole biodiversity of the oceans change? What does that mean? How do we counter that? What would DoD's role be? Geo-engineering, another area we didn't talk much about, could be offered as a solution to global climate change. If climate change turns out to be a really damaging problem and we have to do dramatic things to counter climate change, what's going to happen? Some of these potential geo-engineered changes we don't understand very well. Shooting sulfur particles or aerosols into the atmosphere and building gigantic mirrors up in the atmosphere to block the sunlight; what will be the second and third order effects?[9] What is it going to do to our planet? These could be identified as **irregular** challenges to our national security.

What might we classify as a **disruptive challenge**? Well, what if you have famine[10] on an international scale affecting perhaps four or five regions at the same time? Or what if you have increased fresh water scarcity[11] problems in the countries that are currently suffering from shortages of potable water? What will happen? Also, what if the water shortages spread, for example if the monsoons don't come to India? Climate change also has the potential to influence how diseases spread. How would we respond to a number of different pandemics by different disease vectors? Here we must also consider not just diseases that affect only humans.[12] What will happen if beetles eat all the temperate forests in Siberia?[13] What does that do to Siberia? These challenges run counter to some of our current strengths, and we may have to rely upon revolutionary breakthroughs in technology and science to help us mitigate or adapt to these threats.

Last are the **catastrophic** challenges. Josh Busby said that the probability of the polar ice caps melting is quite low. Still, they *are* melting, and as they continue to melt, sea levels will rise, and bad things will happen.[14] For example, there are several different Ocean Circulation Systems (OCS) which have direct influences on regional and global climate. If the OCSs change, speed up, slow down, or stop, then there will be enormous regional and global changes that could affect crop production, rainfall, fisheries, and a variety of other natural processes that we may not even know about.[15] How do they impact our planet? Finally, is there the possibility of mass extinctions and mass migrations?[16] We didn't talk much about biodiversity, but if you suddenly lose a whole species of organisms within a food chain, what will the impact be? What replaces them, if anything? What will happen? We don't know. Also, if large populations of people can no longer survive in an area or they are displaced by rising sea levels, where will they move to and how will these mass movements of environmental refugees affect international and environmental security? Again, we do not have all the answers.

If any one of these changes takes place, we will face a major challenge. If four or five of them happen at the same time, then we may witness the "perfect storm." In the movie of the same name, a whole series of disparate climatic factors came together to create a super storm, and the storm was impossible for one nation to deal with. Further complicating all these challenges is the fact that natural resources are finite. Is peak oil, the end of fossil fuel abundance, right around the corner; is that coming? And if burning oil is one of the causes of the greenhouse effect and climate change, how do we deal with oil shortages and the impacts of burning fossil fuels? Also, if you throw in the possibility of abrupt climate change, what do we do then? In short, we are looking at a potential problem of no mean proportions.

Solutions.

But I like to view these challenges from an optimistic standpoint. Climate change really offers us a tremendous opportunity. If we focus less on how to deal with the problems that emerge after an event and more about what we can do in advance to prevent problems, help friends, and make new friends, then we might create more good will globally. We have the opportunity to shape our choices and the choices of other states in the future if we approach this in a positive manner with a sustainable security strategy. A sustainable security strategy requires blending the sustainability tenets of environmental security, ecological economics, and social/environmental equity with the pillars of the democratic peace theory.[17] I contend that the conflict ameliorating powers of democracy, economic interdependence, and international organizations operating within the finite environmental, economic, and social limits of the sustainability tenets will enable the United States and DoD to mitigate and adapt to the multiple challenges from climate change and build for the United States and for all other democratic states sustainable security.[18]

Objectives: Sustainable Security.

So suppose we want to prevent the "perfect storm" from taking place. There are two ways you can get at this problem. The DoD has an internal problem and an external problem.

Internally, we in the DoD consume too many finite resources. We are energy hogs. The DoD is the biggest energy consumer in the United States, probably the biggest single energy consumer in the world, if you just wanted to isolate us. We consume way too many resources. And we consume things in unsustainable ways: we don't recycle, we seldom reuse, and we don't have broad based plans to reduce the amount of things we consume. We break things, and then we throw them away.

So, to address all these challenges, the Defense enterprise needs to reform itself. It needs to find a way to create "sustainable security" for the United States. One primary method to create sustainable security is to ensure producers take the product back after the consumer is done with it, and the producer reenters it into the production system and then makes a new one using parts from the old one. I submit that the DoD can become a driving force in the creation of "sustainable security" because it has a lot of leverage. It has a 500 billion dollar budget, which gives it unique purchasing power. Why can't the DoD tell contractors that we want to buy things that have a cradle-to-cradle life cycle?[19] Why can't we put our leverage toward buying those products, buying those technologies that are climate benign, environmentally friendly, and carbon-friendly?

And how do you do that? Well, right now in the climate change discussions, policymakers are talking about two major ways to reduce emissions of greenhouse gases. Economically, they are talking about a global cap and trade system, and they are talking about carbon taxes on emissions. Both of these economic principles are applicable to the DoD.[20] I could see a base commander right now saying, "Hey, we have a carbon tax on our overuse of energy. If we exceed it, then our budget is going to be decreased. Let's find ways to reduce energy use." And everyday, base commanders deal with budget fluctuations. So, we have to reform ourselves.

That was our internal problem. Here is our **external** problem. We focus a lot on how to kill things and break things. We are good at that. We are not so good at making friends. In fact, we should broaden our concept of national security to include social, political, economic, and environmental security. We need to implement ecological accounting and develop a better understanding of how to create sustainable free markets, resolve equity disputes, make democracy work for everyone, and learn how to work with, not against, international organizations to create environmental security. These are the tenets of sustainable security.[21]

Ecological Accounting. There is a sustainability concept called ecological economics.[22] That is where you determine the complete economic cost of your activities and then determine what is cost effective. For example, how much does a gallon of gas really cost? If you figure in all the subsidies, all the money that DoD spends to make sure that oil tankers can get from the Persian Gulf to the United States, one scholar estimated it at about $10.00 a gallon. All the negative externalities have to be included in the true cost of a resource or an activity. Now, figuring in the health costs of burning fossil fuels, how many people are breathing smog—bad air—because of the pollution that comes out of the back end of our cars; figure all those negative externalities in, and the true cost is much higher. Obviously, the real cost of burning fossil fuels is much higher than what we currently see at a gas station or in our electric bill. We have to figure out what the true cost accounting for all DoD processes is. Then we will get an idea of how much our national security efforts are really costing us, and then maybe we can find a better way of doing it, a sustainable way. Then when we compare the complete costs of renewable energy against fossil fuel energy, the fossil fuels are clearly not so attractive.

Environmental Justice/Equity. For those working to mitigate or adapt to climate change, issues of equity are of paramount importance.[23] This word has popped up in four or five of our discussions. We talked about water. Who gets the water? Who controls it? Is it the elites? Is it the people? We talked about land. Who controls the arable land? These are all equity issues that will only be exacerbated by climate change. Also, if you look in-depth into terrorist issues and read the literature on terrorism, you will see that a lot of the problems, a lot of the complaints by people who support the terrorists or actually become terrorists, are equity problems.[24] They feel like they are being marginalized. They don't have a voice. The distribution of life supporting resources is unfair. They don't have anything that they can say to change the system. The system is corrupt, the government is corrupt, and they don't have an avenue to make any positive change. Equity issues come right down to the center of terrorism, but they are also at the center of climate change because the richest and the most developed states are producing the most greenhouse gases. The poorest, most vulnerable, most unequal states are producing the least amount of greenhouse gases. We in the United States have to take the lead in reducing our greenhouse emissions because China and India look at us and say, "You already put all this stuff in the atmosphere. We're not going to do anything until you do something." And you wonder why many developing states won't talk to us? They contend that the United States must make the first move before they are going to believe we are going to make a good faith effort.

Environmental Security. The DoD's existing approach to the natural environment is shallow and unremarkable. DoD policies reflect perceptions of environmental issues more in the realm of pollution prevention, toxic waste cleanup, base closures, and worker safety.[25] What is lacking in the DoD approach is concentrated research into the relationships between environmental/climate change and conflict, and into how environmental security can be sustained. Specifically, Combatant Commanders' Theater Security Cooperation Plans should identify in an Area of Responsibility who controls access to water, food, and energy. Also, the basic environmental context surrounding the water, food, and energy situation must be determined and plans to mitigate or improve basic environmental conditions drafted. These efforts will build trust, cooperation, and goodwill. Additionally, these activities will improve host nation capacity and capability to deal with climate change and other national security threats.[26]

Democracy. Another factor that influences "sustainable security" is governance or simply democracy. This is also a core aspect of our national security strategy.[27] We want to spread democracy. Now *how* we spread democracy is a debatable dilemma. Do you do it at the point of a gun, or do you do it in a more pacific way? That is debatable, but democracy has a certain theoretical appeal that I will discuss shortly.

Sustainable Free Markets. All free markets have to be sustainable and the complete costs of market processes have to be measured and accounted for. We have to make sure that we are not taking away from all these developing countries their core finite natural resources that can not be replaced—if you take away their forests and you destroy or degrade their core natural resources, they will never be able to develop economically, socially, or politically. So we have to find a way that we can deal with these countries so that they can sustain their economies, too.

International organizations. Right now, we don't do a very good job of working with governmental and nongovernmental organizations (NGOs). We do a really good job at heading them off and stopping them from accomplishing things we don't like, but we don't work with them as well as we could. So this is the core of our "sustainable security" strategy.

Developing a 21st Century Total Force.

The QDR says we need to reconfigure the total force, build the right skills, and effectively use our human capital.[28] I think these concepts can aptly be applied to the challenges created by climate change. So what would a sustainable security force developed to deal with climate change look like as opposed to what we currently have? Actually it would look a great deal like the forces we currently have. You could use the QDR as it is, merely changing a few words here and there, and it would be a good starting point for creating sustainable security forces.

A new sustainable security force would be focused on what climate change is doing internationally to countries that are vulnerable to global warming. The new force would be a mix of military and civilian specialists who understand what climate change could do to make these states unstable, and what security issues might potentially create problems for us. The sustainable security forces would be expected to head these problems off before they blend together in a "perfect storm" and create failed states and breeding grounds for terrorists. And, as you may remember, during the tsunami event in 2005 when the Navy showed up on ships with helicopters, food, water, and shelters, we were very popular in South East Asia again. In particular, our popularity rating in Indonesia soared from that one event. Those types of events, where we intervene to help states mitigate and adapt to climate change and then leave, will build trust and confidence in the United States. If we do this on a recurring basis aided by our allies, maybe we can gain back that global goodwill we used to have.

Unity of Effort.

The QDR also noted that we should try to achieve **unity of effort**.[29] This included a suggestion that we do things to make the interagency process work better. I will leave that aside for the moment. Another QDR recommendation was that we improve our abilities to work with **allies and partners**.[30] We haven't done a good job lately of working with allies and partners on a recurring basis. And we have to do a better job at getting out the strategic communication, which is at the heart of countering terrorism—telling the people that "Hey, we are selling something better than what the terrorists are selling." Strategic communication clearly has climate change applications. We need to tell vulnerable states that we will help them mitigate and adapt to climate change. We need to build their capacity to deal with climate change, and then their internal security will be increased. This will make us more friends and allies.

Within the political science field, scholars have developed something called the democratic peace theory. The three key processes that make the theory acceptable are the interactions of democracy, economic interdependence, and international organizations.

They work together, and they create international peace. That is the theory. And believe it or not, it has been proven in a variety of ways, and it is working. Democracies don't fight each other.[31]

Now, how do you get more peace and sustain it? Well, as I said before, you take some of the core paradigms out of the sustainability argument: environmental security, ecological economics, and that idea of equity, and you try to push them inside of the democratic peace theory into what scholars call the "pillars of the democratic theory." There are a couple of well-known political scientists, Bruce Russett and John O'Neal, who call the relationship between the three pillars a "virtuous circle."[32] It is based on the famous philosopher Emmanuel Kant's theory of perpetual peace and how these processes and institutions work together to reinforce each other and create peace.[33]

What I recommend is that democracy be made the core governmental structure of any sustainability process. Take ecological economics and embed it into the processes that are creating economic interdependence so that our economic processes become sustainable. Finally, take our international organizations and use them as the tools to spread and implement sustainability principles, with a clear focus on creating equitable living conditions globally. Give those voices out there that have not had a chance to speak an opportunity to be heard. This will green the democratic peace theory and will hopefully foster sustainable security.[34]

Why the DoD?

Now why should DoD play a role in all this? What is unique about DoD? DoD has many positive characteristics that can allow us to use it as a starting point to implement these ideas. First, one of the main objectives of our national security strategy is to help create more democracies around the world. Also, our charge is to spread prosperity and freedom and to protect human dignity,[35] which I would equate to creating an equitable world where people have a chance to succeed. That is our charge in DoD based on the objectives identified in the National Security Strategy.

Also, the DoD has got the biggest budget in the U.S. Government. There is that leverage tool again. Now, if you take that big budget and you make it sustainable and you apply ecological economic principles to the way we buy and sell things, preferably buy things, then we will be more sustainable and the DoD procurement process will be greener. And we have done some good things in DoD in the past, but we have also dirtied up some places. We dumped some rather toxic things and buried dangerous things and some of them, we don't even know where they are. We have to do a better job. If we don't remember where we put some of these toxic things, they are going to come back some day and create problems. They are going to end up in the ground water, or the radiation from some of these dumps is going to travel. You have to be careful what you do with some of these things.

In addition, DoD has been a social laboratory in the past. For example, DoD was a pioneer for desegregation, minorities working in the military, religious freedom, and preventing sexual harassment. All those issues have been addressed in the military early on, and I think we have done a pretty good job of overcoming many of those challenges.

Finally, there is one more major challenge—social

and environmental equity issues. How do you go into somebody else's country where less than 10 percent of the people control 90 percent of the land? Essentially, the land in many parts of the developing world is controlled by a small group of elites. How do we get them to give up that power and spread the wealth and the land? And, we use too much energy; we are the energy hogs out there. We use more energy than anybody else, so if we can move from being fossil fuel dependent to being renewable based and sustainably based, then we can set the example for the rest of the federal agencies and also for the state governments and for other countries. Also, DoD owns a lot of land, so that gives us a lot of opportunities to show how to use our natural resources in a sustainable manner. New technologies are coming out where we can sequester the carbon dioxide emissions and reforest areas and things like that to take more carbon out of the atmosphere.

The bottom line is we must take the lead and use the pacific forces of democracy, united with economic interdependence based on ecological economic principles. We should encourage and enable international organizations to implement equitable climate change regimes and treaties. We must also lead the world in securing perishable environmental areas that are threatened by climate change and help states create sustainable environmental security within their own borders. If we reach out to other nations and try to help them mitigate or adapt to climate change, we will undermine the equity arguments that terrorists take advantage of, we will make it less likely that climate-induced failed states will become breeding grounds for extremists, and we will return the United States to the position of global friend and leader, admired the world over for leading the efforts to prevent catastrophic climate change!

ENDNOTES - Ackerman

1. Department of Defense, *Quadrennial Defense Review Report* (QDR), Washington DC: Office of the Secretary of Defense, 2006, p. 19.

2. *Ibid.*

3. Peter Schwartz and Doug Randall, "An Abrupt Climate Change Scenario and Its Implications for United States National Security," Emeryville, CA: Global Business, 2003, available from *www.gbn.com/ArticleDisplayServlet.srv?aid=26231.*

4. Gregory D. Foster, "A New Security Paradigm," *World Watch*, January-February 2005, p. 39; and John T. Ackerman, *Security and Sustainability: A Critical Analysis*, Colorado Springs, CO: U.S. Air Force Institute for National Security Studies, 2007, p. 82.

5. We did deal much earlier in this conference with the major roots of climate change which are the burning of fossil fuels and wide-spread regional land changes.

6. IPCC, A Report of Working Group II (IPCC WG II), *Climate Change 2007: Climate Change Impacts, Adaptation, and Vulnerability*, World Meteorological Organization and United Nations Environment Programme, 2007, pp. 4-7.

7. *Ibid.*, p. 8.

8. IPCC, A Report of Working Group I (IPCC WG I), *Climate Change 2007: The Physical Science Basis-Summary for Policymakers*, World Meteorological Organization and United Nations Environment Programme, 2007, p. 2.

9. James R. Fleming, "The Climate Engineers: Playing God to Save the Planet," *The Wilson Quarterly*, Vol. 31, No. 2, Spring 2007, pp. 36-46.

10. IPCC WG II, p. 8.

11. *Ibid.*, p. 7; Scientific Expert Group on Climate Change (SEG), Rosina M. Bierbaum, John P. Holdren, Michael C. MacCracken, Richard H. Moss, and Peter H. Rave, eds., *Confronting Climate Change: Avoiding the Unmanageable and Managing the Unavoidable*, report prepared for the United Nations Commission on Sustainable Development, Sigma Xi, Research Triangle, ND, and the United Nations Foundation, Washington DC, 2007, p. 22.

12. IPCC, A Report of Working Group III (IPCC WG III), *Climate Change 2007: Mitigation of Climate Change*, World Meteorological Organization and United Nations Environment Programme, 2007, p. 11.

13. National Assessment Synthesis Team (NAST), *Climate Change Impacts on the United States: The Potential Consequences of Climate Variability and Change*, U.S. Global Change Research Program, Washington, DC, Cambridge University Press, 2000, p. 77, available from *www.usgcrp.gov/usgcrp/nacc/default.htm*.

14. SEG, p. 20; Eric Rignot and Pannir Kanagaratnam, "Changes in the Velocity Structure of the Greenland Ice Sheet," *Science*, Vol. 311, No. 5763, 2006, p. 986; Jonathan T. Overpeck, Bette L. Otto-Bliesner, Gifford H. Miller, Daniel R. Muhs, Richard B. Alley, and Jeffrey T. Kiehl, "Paleoclimatic Evidence for Future Ice-Sheet Instability and Rapid Sea-Level Rise," *Science*, Vol. 311, No. 5768, 2006, pp. 1747-1748; IPCC WG II, p. 17; Arctic Climate Impact Assessment (ACIA), *Impacts of a Warming Arctic: Arctic Climate Impact Assessment*, Cambridge, UK: Cambridge University Press, 2004, p. 94.

15. IPCC WG II, p. 17.

16. Camille Parmesan, "Ecological and Evolutionary Response to Recent Climate Change," *Annual Review of Ecology, Evolution, and Systematics*, Vol. 37, 2006, p. 657; SEG, pp. 24, 26; IPCC WG II, p. 8.

17. Bruce Russett and John R. Oneal, *Triangulating Peace: Democracy, Interdependence, and International Organizations*, New York: W. W. Norton & Company, 2001.

18. Ackerman, pp. 87-97.

19. William McDonough and Michael Braungart, *Cradle to Cradle: Remaking the Way We Make Things*, New York: North Point Press, 2003.

20. Lester Brown, *Plan B: Rescuing a Planet Under Stress and a Civilization in Trouble*, New York: W. W. Norton & Company, 2003, pp. 233-235; SEG, pp. 136-137; Nicholas Stern, "Stern Review: The Economics of Climate Change," Executive Summary, Cambridge, UK: Cambridge University Press, 2006, p. xvi; Center for Naval Analyses (CNA) Corporation, *National Security and the Threat of Climate Change*, Alexandria, VA: The CNA Corporation, 2007, pp. 39, 47-48.

21. Ackerman, pp. 87-97.

22. See, for example, Brown, *Plan B*, 2003; Herman E. Daly, "Economics in a Full World," *Scientific American*, Vol. 293, No. 3, September 2005, pp. 100-107; Robert Costanza, *et al.*, *An Introduction to Ecological Economics*, Boca Raton, FL: St. Lucie Press, 1997; Robert Costanza, *et al.*, "Managing Our Environmental Portfolio," *BioScience*, Vol. 50, No. 2, 2000, pp. 149-155; Robert Costanza, "Visions, Values, Valuation, and the Need for an Ecological Economics," *BioScience*, Vol. 51, No. 6, 2001, p. 464.

23. See, for example, Eileen Claussen and Lisa McNeilly, *Equity and Global Climate Change: The Complex Elements of Global Fairness*, Arlington, VA: Pew Center on Global Climate Change, June 2000; Urs Luterbacher and Detlef F. Sprinz, *International Relations and Global Climate Change*, Cambridge, MA: MIT Press, 2001; Anil Agarwal, "A Southern Perspective on Curbing Global Climate Change," in Stephen H. Schneider, Armin Rosencranz, and John O. Niles, eds., *Climate Change Policy: A Survey*, Washington DC: Island Press, 2002; IPCC WG III.

24. Paul J. Smith, "Climate Change, Weak States and the 'War on Terrorism' in South and Southeast Asia," *Contemporary Southeast Asia*, Vol. 29, No. 2, 2007, pp. 264-285.

25. Ackerman, p. 82.

26. CNA, p. 40.

27. George W. Bush, *The National Security Strategy of the United States of America* (NSS), Washington DC: The White House, 2006.

28. QDR, pp. 75-82.

29. *Ibid.*, pp. 83-92.

30. *Ibid.*, pp. 87-91.

31. Russett and O'Neal.

32. *Ibid.*

33. Emmanuel Kant, *Basic Writings of Kant*, Introduction by Allen W. Wood, ed., New York: Random House, 2001; Russet and O'Neal, pp. 35-42.

34. Ackerman, pp. 87-97.

35. Bush, p. ii.

Commentator

Col. Gregory M. Douquet, USMC
Royal College of Defence Studies
and King's College, London

As a practitioner, currently in the U.S. Marine Corps, I, too, have been wrestling with how, and indeed if, we should establish linkages between U.S. National Defense and climate change, which is the question put before us on our panel today. I would like to thank our speakers, Kent, Joshua, and John, for what I thought was an excellent analysis of a very complicated and important issue.

It struck me that all three speakers agreed on some issues, disagreed on others, and there were a few issues that they agreed upon the nature of the problem but there were differences in the way they thought the U.S. Government should approach it. Incidentally, the internal debate that we are having here I find an interesting contrast to the debate in the United Kingdom (UK), where the current Defense Minister, John Reid, was recently quoted as saying that "climate change is the single biggest threat to UK national security, indeed, greater than international terrorism."

So, what is it that all three speakers agree upon? Well, first of all, they agree that the climate is changing, and this, at a minimum, will complicate U.S. national security and defense strategy and possibly create situations in which the United States might choose to intervene. They also agree that if the United States responds to a climate change induced problem, the response should be via the interagency that coordinates and employs the appropriate instruments of national power.

Perhaps in the question and answer (Q&A) period, we can discuss if such a situation does demand an interagency response, is our current interagency up to the job? They also agree that climate change might be a way to help restore U.S. soft power that might be sorely lacking in the world today. Our three speakers saw some climate change issues differently, however.

For example, how we might establish conditions to counter the effects of global climate change and the linkages between national defense and climate change? Joshua saw the need for a plausibility probe, and Kent and John saw climate change as a more or less imminent threat, something that we need to think about now and perhaps take action on. And all three agree that national defense needs to be part of the U.S. Government's response to global climate change, and that climate change could be part of a larger problem such as complicating our struggle against extremist ideology.

There were some differences in how they think the U.S. Government should approach the problem. For example, Kent suggested the military can shape the environment by regional engagement via the Combatant Commander. Joshua thought that we could focus upon climate change with regard to extreme weather events. Perhaps we can probe these differences a little bit further in the Q&A. Each of our speakers offered some unique perspectives on the issue of U.S. national defense and climate change, perspectives I have not heard before.

I thought that Joshua's analysis of the conditions under which climate change can pose a threat to U.S. national security interests was insightful and helps focus our attention, and maybe more important in these days of stretched capabilities, our resources.

As Joshua pointed out, many of the effects of climate change are beyond the scope of national security, and perhaps we need a new model to address them.

Moreover, it is hard or probably a misallocation of resources to prematurely redirect national security strategy or national military strategy in expectation of theoretical climate changes while we are fighting a long multifront war, and our forces are deployed around the world. I particularly agree with Kent's observation that environmental security issues often trigger or exacerbate conflict and with his idea of Maslow's Hierarchy of Needs: that environmental security is often not thought of as important by developing countries, yet it is the foundation of their very success; we have heard a lot about that today.

We should remain mindful of what John said about sustainable security, and the fact that the DoD is the biggest user of energy in this country and made perhaps a great step in the right direction with respect to the demand side of the equation.

I would like to make a few observations from a practitioner's perspective. In terms of national defense, climate change is likely to make a bad region worse. Those regions destabilized by resource shortfalls and great ethnic tensions, disease, and poor governance, the stresses that we have been discussing all day today, could push these regions over the edge through the exacerbating influence of climate change. Sub-Saharan Africa and Central and Southeast Asia come to mind. Therefore, it seems to me that a regional approach, as opposed to a DoD-wide, one size fits all approach, is reasonable and feasible. We have a mechanism for such a regional approach: Unified Combatant Commanders. They are among our eyes and ears around the world and are probably in the best position to assess the affects of climate change and work with their regional partners, military to military and state to state, to shape the situation before it erupts into a larger problem.

But, "What might these Combatant Commanders do?" is the question, particularly when these Combatant Commanders don't own their own resources, their own forces. They are assigned those forces and as of right now, none of those forces are assigned for the purpose of responding to climate change. Perhaps they could do some of the eyes and ears aspect of their job. Risk management and modeling has come up today. Perhaps they could be a means for nations to openly cooperate, military to military. Perhaps they could think of pre-staging capabilities, in order to respond more rapidly in a climate change type of situation.

Discussion

Q. (Douquet): If I may, I'd like to steal the first question for our panel and ask it about the possibility of U.S. intervention in a region destabilized by climate change. It strikes me that such a projected intervention is something of a Catch 22. Darfur might be a good example. If the United States does intervene, regional powers will likely be suspicious of intentions and portray U.S. actions in a negative light. It will also give those who want to gain an advantage over the United States an opportunity to criticize us, possibly also decreasing our soft power.

If the United States doesn't intervene, we will be accused of having created the problem with our carbon emissions, *et al.*, what John was talking about, and not participating in the solution. In what might be perceived as a sovereign nation's domestic problem, what would be our justification to intervene and would such an intervention likely meet UN approval?

Busby: My first reaction would be to look at the part of the world where it took place and then think about the strategic significance. If we are talking about Africa, I think its strategic importance to the United States is an open question. Looking at what our interests are in the region comes down to resources, the possibility of state failure creating ungoverned spaces for terrorists to organize, and the potential for blowback if we fail to intervene. Even given those considerations, much of Africa may not be strategically important to the United States. So intervention would have to be justified on moral grounds, which is a hard conclusion to come to as someone who is concerned about Africa. But merely thinking about how extreme weather events might exacerbate human tragedies is an important cognitive shift on our part. We also see the mobilization of domestic constituencies that demand the U.S. military take action even when we have no strategic dog in the fight. We should anticipate this possibility. Investing in early warning systems to head off such crises will make it less costly for us later on.

Butts: I agree. I like to approach intervention like lifesaving: We should row, throw, and then go. Go last. Intervention should be a last resort. Have we given diplomacy a chance? Are there others that could participate? Have we tried our alliances? What about the UN? Could we, in fact, support an international force by providing strategic lift and intelligence, but not having our troops on the ground themselves? There is a reason why you don't use the most powerful military in the world in a cavalier fashion. When you do get involved, you should be sensitive to local customs and have a game plan. That way you won't create resentment. There are times, however, when we should get involved in a place even if it is not strategically important to us. If you look at the Great

Lakes area in Rwanda, many people would say that the number of dead could have been halved if the United States had been involved earlier. It is hard to walk away from that. I don't think we should. Finally, if we do get involved, we should build the necessary institutions and capacity to do a hand off. The World Bank used to say "The problem with U.S. military is you get in, then because you don't want mission creep, you get out as soon as you can, and then we have a potential failed state." So while you are in there, create those institutions. Work with NGOs to build capacity so that they can be effective when in fact we pull out and maybe allow us to pull out earlier.

Ackerman: I would like to differ a little bit from my colleagues. I think intervention is an opportunity, especially if it is done in response to a disaster that is climatically or environmentally driven. I think that in such cases, we don't have to worry so much about countries construing our intentions in a negative way. When we responded to the tsunami, we generated a great amount of good will. I think that we ought to take advantage of any opportunity out there to help countries mitigate or adapt to climate change. Stability, Security, Transition, and Reconstruction (SSTR) operations are designed to help countries develop the ability to solve their own problems using their own security forces. We help them understand the climatic problems that they are going to encounter in the future and build their own capabilities. So I think helping a country deal with an unnatural disaster is an opportunity.

Q: My concern in that you are calling upon DoD to take on a lot responsibilities that are more properly suited for the DoS. Could you comment on that, please?

Ackerman: The DoD has a 500 billion dollar budget, the largest in the United States Government, and other necessary capabilities and assets. The DoS has nothing compared to this. Moreover, there has been a transition in the last 4 or 5 years to put even more capacity into DoD and take it away from DoS. So as long as that trend continues, I think that DoD is the logical organization to turn to. That is not to say that DoS cannot provide cultural and ecological awareness and also help with those transitions and security situations. But DoD has the power; DoS does not.

Audience: But DoS is the lead for foreign disaster and assistance. DoD cannot deploy forces without DoS support. A Combatant Commander can take the initiative in the theater and introduce programs of humanitarian assistance or disaster preparedness—as in the case of Indonesia. However, when it comes to providing a system of enduring presence and enduring assistance after the disaster, that would be better done by interagencies and other governmental agencies. The military likes to fight its nation's wars. We are not the world's 9-1-1 emergency carte blanche, and I think we all understand that.

Busby: The real problem is that we do not have adequate tools in the State Department tool kit in an expeditionary way to do nation-building or provide adequate state governance assistance. We have ended up with a concentration of capability in the Pentagon, and we see the atrophying of USAID. I agree that, at the end of the day, the military is going to be tasked to fight and win wars; but when we face new challenges, we need to anticipate and build the capacity to deal with them. And whether those capabilities lie in Defense or State or some other venue, we should ensure that military capabilities aren't the only tool in our arsenal.

Q: I have a question about priorities. The British Government spokespersons have said that, in their view, climate change is the greatest threat to their national security. We are in a country where that is not the view of our administration. The panelists made some very thoughtful recommendations as to what we should do to meet the coming threat. But what is the likelihood that any commander is going to take these steps? The military establishment has to meet a lot of other operational and other requirements. Given that climate change is viewed as a long-term rather than an immediate problem in many quarters, including our own administration, are we going to see anything happen in the foreseeable future along the lines that you all are suggesting?

Butts: The answer is "Yes." As General Frank said, "It doesn't matter that we are in two wars, we still have to do engagement." Not so long ago, CENTCOM was told by the Deputy Assistant Secretary of Defense to abandon its environmental security programs despite the fact that Deputy CENTCOM Commander General Delong said, "If it weren't for those environmental security programs with the Central Asian States, it would've taken us years instead of months to gain the access we needed to fight the Afghanistan War." But CENTCOM went ahead anyway – they simply renamed the program. Three weeks ago we helped CENTCOM do a nuclear disaster preparedness conference for the GCC plus 2, Jordan and Egypt. Despite the pressures, they went ahead with their engagement program. This is true of other commands, too.

Dorff: The Deputy Commandant Pacific Command and European Command are, I can confirm, involved in engagement. They are involved because in Washington there is a lot of talk but little action and because they know it must be done. Out there in the theater, things are being done.

Q: You suggested that, because DoD has the largest budget, it should be tasked with humanitarian intervention. This same logic might be applied to everything else that the government does, as well. Is this not a case of mission creep?

Ackerman: It would indeed be a problem if we saw the DoD as a bottomless pit and sought all our answers from it. I did not mean to suggest this. I do think, though, that we will achieve positive results if we help people deal with climate change. Suppose a third world nation suffers from serious floods or the kind of heat wave that causes security to break down in cities. We could use our soft forces to help them adapt to these challenges. We might be fondly remembered.

Q: Let me ask a hypothetical question. What if a disaster on the order of the tsunami occurred in Iran? It would seem to me that if we offered to help them, they would see that as an opportunity for us to affect regime change. They wouldn't want us in there because the U.S. military, for whatever reasons, good or bad, has a reputation of not simply doing humanitarian relief. Rather, it tries at the same time to achieve larger government objectives, which do include in some places like Iran, regime change. How would you respond to a situation in which there was an opportunity to intervene to affect humanitarian relief, but in which there was a larger political framework, which might, to put it mildly, complicate the achievement or intervention?

Ackerman: After Hurricane Katrina, we received an offer from the Iranian government and also from some other governments that we are not very friendly with to aid us, too. That was rather unique, I thought. As to the Iranians, they might not want our help, and we would obviously not force it on them. But it would surely say a lot about us if we offered to help people we weren't on very friendly terms with. It might begin to help us recreate the idea that America is about democracy and freedom and human dignity. I would hope, though, that whenever we go, we go with the UN and we go with NATO (North Atlantic Treaty Organization), we go with the European Union (EU), we go with a whole group of allies.

Q: Will climate change oblige us to think in unfamiliar ways? Or will we be dealing with things we already understand?

Busby: There will be some new things—glacier melt in Greenland and Antarctica that could surprise us, for example. There may also be major discontinuities of the kinds that we heard about where there is larger ecosystem collapse that we may have difficulty anticipating. But some things over which we have more confidence are the increased severity and frequency of extreme weather events that we are more familiar with. But we haven't really thought about and prepared for how to handle the socio-economic and political consequences of them.

Ackerman: Global agreement enforcement may be challenging. Suppose, for example, we sign a treaty agreeing to put a cap on oil. How do we stop the illicit trade in oil? Would we use the U.S. Navy to police the world?

Glantz: I cannot see the military policing the oceans. We are not a bottomless pit. As my periodontist said when he gave me a toothbrush, "Brush the teeth you want to keep." We need to think about our priorities. If we want to protect the world, let's do it. If we want to protect certain strategic interests, let's do that. But we do need to deal with problems at home, too. Our failure to deal well with Hurricane Katrina at home has undermined foreign confidence. In Spain, our allies wonder if they can depend on a nation that can't take care of its own! We also need a reality check. A lot of what we are talking about here is theoretical, and we can do anything we want on paper. Every early warning system that I know of works on paper and in Power Point. But in reality, they all don't seem to do the job. I would like to see more discussion about reality.

Butts: We should remember the CNN factor. Take, for example, the Horn of Africa and Somalia. You could not have paid any military person to bet a nickel that we would have put 20,000 combat troops in the Horn of Africa to distribute food. And yet we did. And we could be back tomorrow, depending on what is shown on television. Public opinion is very powerful, and so is the media.

Busby: In my paper, I tried to distinguish between things that were more and less credible. I focused on extreme weather events because these, I think, are more likely to command the attention of policymakers in the short run. It may be that we ought to have the broader strategic discussion in the lead up to the 2008 elections. But at the very least, we are injecting into the discussion a new issue. Concerns about climate change do have some scientific basis and we ought to be paying them some attention.

CHAPTER 4

MEETING THE CHALLENGE

In this chapter, discussion shifts from analysis of the problem to a discussion of solutions. What individuals/ institutions should be involved in the effort to combat climate change? What methods are most likely to yield positive results? Mr. Dennis Tänzler stresses the need for a transatlantic consensus and effective international diplomacy, Mr. Simon Rich focuses on the role of corporations, and Dr. Anthony Leiserowitz looks at risk perception and analyzes how this affects our ability to deal with the issue.

International Diplomacy

Dennis Tänzler
Adelphi Research

My talk today will focus on the mitigation challenges facing the transatlantic and the global communities. In particular, it will examine the prospects of transatlantic climate and energy policymaking. There are good reasons to think about these issues. Science tells us that we have only a limited window of opportunity.[1] If we do not take decisive action within the next 5 to 10 years, it will be hard to avoid some of the worst impacts of global climate change.

In my presentation, I will touch on four points. First I will outline the milestones of international climate policies. Second, I will argue that the basis for a future global climate agreement will be a transatlantic consensus. I want to explain why I think there is common ground, and why I am optimistic that we will be able to start developing forward looking climate and energy policies. Third, I will discuss ways to engage other countries and suggest what issues need to be part of a global framework on climate change. Fourth, I will suggest ways to inject some new dynamics into the international climate-change process.

Milestones of International Climate Policies.

When we talk about the milestones of international climate change policies, there are basically two major international agreements; a very strong United Nations (UN) framework convention on climate change agreed on in 1992, and the Kyoto Protocol, adopted in 1997.[2]

1991 United Nations Framework Convention. The 1992 UN framework convention was ratified by 189 countries. Its ultimate objective is to avoid dangerous climate change. Some analysts have argued that it is more or less a security treaty for those countries (especially small island states) which are in danger of disappearing because of rising sea levels. An important normative aspect of the framework convention on climate change is that it points out the common but differentiated responsibility of industrialized and developing countries to combat climate change. In other words, it works on the assumption that industrialized countries, because of their historical responsibility for causing greenhouse gas emissions, should take the lead in reducing greenhouse gases and the developing countries will follow.

1997 Kyoto Protocol. The Kyoto Protocol was adopted in 1997 after "negotiation by exhaustion." This protocol has been subject to highly controversial debates within the transatlantic as well as international community. It did not enter into force before 2005. It has been ratified by 169 countries but not by the United

States and Australia. The Kyoto Protocol includes legally binding emissions reduction targets for industrialized countries. It does not set emissions reduction targets for developing countries. It does introduce some flexible mechanisms including emissions trading which are designed to promote sustainable development and reduce overall emissions.

The idea behind emissions trading is that countries who have come in under their prescribed limits can sell the emissions permitted but not used to countries who have failed to meet their limits. Countries who have to pay a steep price to "buy" compliance will be encouraged to promote the development of alternative sources of energy.

This system means jointly implemented emissions reduction projects within the framework of the Clean Development Mechanism (CDM) or Joint Implementation (JI). The CDM is a good example of international cooperation. Industrialized countries invest in projects in developing countries that will help reduce emissions. By doing so, they get so-called certified emission reduction units which means they will not have to reduce emissions so much in their own countries. Such certified reduction units or emission permits might, I think, become a powerful world currency.

However, these innovative instruments are—at least in the way they are currently designed—not sufficient to trigger breakthrough technological innovations. There are no appropriate mechanisms or agreements to enforce technological change. This might well be the central reason why industrialized countries today do not perform very well when it comes to emissions reductions.

Copenhagen: Home of a Future Agreement? Nevertheless, plans are underway in the international community to start negotiating a new climate agreement by the end of the year.[3] There is some urgency here if we are to reach an agreement by 2009 on what a 2012 climate change framework should look like. The conference will take place in Copenhagen, and the agreement is thus likely to be called the Copenhagen Protocol or Copenhagen Agreement. The negotiations, as well as the agreement, will tackle the questions of mitigation, adaptation, financing, and technology.

A main feature of this agreement is likely to be differentiated commitments. Countries that contribute most to global greenhouse gas (GHG) emissions will be asked to reduce GHG emissions substantially in the years to come. We need some kind of technology funding mechanisms to trigger innovations and, of course, we will have to raise the question of adaptation and compensation. Technology funding mechanisms might be an appropriate way to link all those four challenges.

It is clear that these negotiations will be very complicated. Creative solutions and willingness to compromise are prerequisites for a successful outcome. Four components will be decisive. First, the European Union (EU) needs to demonstrate renewed leadership. That means the EU has to show that the climate policies it has adopted so far are successful. Second, we need to foster U.S. leadership. Third, we need to engage emerging countries like China and India which today belong to the major emitters. And, last but not least, the international community needs to compensate countries most affected by climate change.

The Need for a Transatlantic Consensus.

If these negotiations are to be successful, a transatlantic consensus is crucial. Why?[4] First, the United States, Canada, and the EU are responsible for about two-thirds of industrialized GHG emissions. Between them, they caused the lion's share of the anthropogenic greenhouse gases in today's atmosphere. Because of this historical responsibility, emerging economies like China and India are only likely to follow and to accept legally binding commitments if the United States and the EU take the lead in reducing greenhouse gas emissions.

Second, transatlantic political and economic cooperation has a vital role to play in encouraging innovations and curbing emissions. Forward looking policy measures on both sides of the Atlantic will do much to encourage environmental innovations. The increased use made of renewable energy in Germany and wind energy in Texas are cases in point.

At the same time, decisionmakers on both sides of the Atlantic should be aware that their first and foremost obligation is to stop and reverse negative emissions trends. The GHG emissions trends from 1990 to 2010 make it quite clear that policy efforts so far have been insufficient. Two trends are visible. On the one hand, there has been a tremendous increase in U.S. GHG emissions since 1990, although data for 2004 and 2005 indicates that this negative trend may at least have slowed down.[5] On the other hand, the EU has shown since 1990 that a reversal of the emissions trend is possible. However, the EU is still far away from the emissions reductions it aims to achieve. It still has some way to go before it is in compliance with the Kyoto target.

On a more optimistic note, there is some reason to argue that the EU will achieve the target since some of the measures adopted will show impact only in the years to come. One important example in this regard is the development and implementation of emissions trading systems. Emissions trading was invented initially in the United States. It is now a major pillar of European Climate Change Policies and provides us with an example of what a transatlantic climate policy approach might look like.[6]

Prospects for a Transatlantic Consensus.

There are several reasons why I think such a consensus may be reached.

Scientific Certainty. First, the dialogue between the United States and the EU countries is based on increased scientific certainty. If we compare today's discussion regarding human influence on the climate system with that 5 years ago, I think we see a tremendous difference.

Converging Climate and Energy Security Agendas. Second, climate and energy security agendas are beginning to come together on both sides of the Atlantic.[7] In **Europe**, in March 2007 there was a very important EU council meeting in which EU heads of state agreed upon several binding targets.[8] They agreed that by 2020, 20 percent of energy used must be renewable energy and 10 percent of EU transport fuel must be bio-fuels.[9] They also made a commitment to improve energy efficiency by 20 percent in the same year. This is important because, as the European Commission pointed out in the fall of 2006, energy efficiency is one way to significantly reduce greenhouse gas at minimal cost. A number of observers regard this

EU council as a successful integration of both agendas, that of climate change and energy security. Some are so euphoric as to see in the meeting the rebirth of the European spirit.

Meantime, in the **United States**, there are also encouraging signs of a similar convergence of climate and energy security agendas. For example, in the State of the Union Address of 2007, President Bush pointed out that climate change is an important issue. He also said that there is a need to increase energy independence and cut gasoline usage by 20 percent in 10 years. He mandated a higher proportion of alternative fuels and announced measures for increasing the fuel efficiency standards for cars and light trucks.

Even more importantly, there are promising activities at the state level. For example, more than 20 states have introduced renewable energy portfolio standards. It is worth noting why these states agreed to adopt binding targets regarding their share of renewable energies. Analysis suggests two reasons: States are guided by environmental policies, and they recognize the need for energy independence.[10]

Cutting GHG Emissions. There are other promising signs of a transatlantic consensus—in this case reflected in a mutual interest in reducing dangerous emissions. In **Europe**, at a March 2007 meeting, EU members agreed on a binding unilateral commitment to cut 1990 levels of GHG emissions by 20% by 2020. In addition, the EU emphasized that developed countries need to reduce their own GHC emissions by 60 to 80 percent by 2050.

In the United States, there is huge political debate as to whether to set a cap on GHG emissions. In the U.S. Senate, a number of proposals for a cap and trade system were introduced in 2007. These included proposals for caps to be set in 2020 and 2050, respectively. Moreover, regional climate policy approaches in California aim at similar long-term emissions reduction targets. In other words, EU and the United States at least have a place to start when it comes to discussing what a future climate change policy should look like.

This is especially the case with respect to the development and implementation of emissions trading systems. The EU will leave the learning phase of its system in 2008.[11] Meanwhile, regional systems are under development in the United States. Here I draw your attention to systems developed by nine northeastern states—the Regional Greenhouse Gas Initiative (RGGI)—and a system planned by California and five other west coast states.[12]

The federal system of the United States makes it likely that these approaches at the state level will have a significant effect at the national level. Pluralistic competition means that policy innovations like standards and procedures for emissions inventory and registry are likely to spread across the whole country after a "critical mass" has been reached. In addition, there are senate proposals for national cap and trade systems to be discussed in the upcoming months.[13] Most importantly, we need to mention the Lieberman proposal which is co-sponsored by potential future presidents of the United States John McCain, Barak Obama, and Hillary Clinton. Given their ambitions, their political support means something for the development and implementation of a national emissions trading scheme.

What might be as important as the political will I outlined is the increasing pressure by business actors to establish market-based solutions to tackle the problem of global climate change. A number of important companies and business have asked for a global market for emissions in order to get investment security and the same conditions in the United States and within the EU.[14]

Engaging the International Community.

So the United States and Europe have both made progress when it comes to these issues. They share common ground when it comes to such things as the need to cut greenhouse gases and the convergence of climate change and energy security issues. They have much to learn from each other, and both could and should work together.

Extending the Partnership. There is another reason to stress the importance of the transatlantic relationship. A cooperative relationship between the United States and the EU could serve as a catalyst for broader global action involving newly industrialized countries. Given the increasing emissions in countries like China and India, engaging other countries is vital. What must we do to bring such nations into the process?

Transforming Energy Systems. First, we need to transform energy systems. We need to achieve a global deal on low carbon technologies which can lead the way towards a decarbonized future. This must be achieved by the middle of the century at the latest. In particular, we need to increase efficiency in the power plant sector. We need to improve our ability to capture and store carbon and develop better gasification technologies. We need to expand renewable energies, and to develop high-capacity grids. Finally, we should not forget that from a global perspective, decentralized energy systems can play a very important role in alleviating energy poverty throughout the world.

Flexible Emissions. Looking at the debate within the international climate change arena, there are signs that newly industrialized countries are more likely to agree on a first step to a flexible or a sectoral emissions target.[15] However, we should figure out what our options are when it comes to linking targets and gaining improved access to clean technologies. This is especially necessary given the huge increase in power consumption in China and India. We also need to develop a system of adaptation planning and financing which needs to start immediately. There are some funds under the UN framework convention on climate change and under the Kyoto Protocol, but they are underfunded and not well equipped. So we need to gain new momentum here.

Visionary Projects. Another related approach is to develop visionary projects, or lighthouse projects, in order to jointly address mitigation and adaptation needs. Here I may point to the discussions within the EU on an Africa-EU energy partnership. At the same time, new concepts are being developed in the area of renewable energies based on solar thermal power plants like those already operating in Kramer Junction, California. How, for example, might one build a transmission grid to connect Europe, the Middle East, and North Africa, the so called EU-MENA region? The objective is to provide clean energy for the sunbelt countries as well as for Europe. The vision is to develop a system of solar thermal power plants in the desert that would serve multiple purposes: reduce $C0^2$ emissions, increase the security of energy supplies in EU-MENA, and provide additional jobs and earnings for people in North Africa and the Middle East.

This concept was suggested by the National Energy Research Center of Jordan, together with the Club of Rome and the German Aerospace Center.[16] They developed a concept based on solar thermal power plants connected through an intercontinental electricity transmission grid with the European grid. This grid would enable the EU to obtain about 10 percent of its electricity demand from MENA countries. In addition, there are plans to use the waste heat generated by the plants to desalinate water and hence to tackle the problem of water scarcity at the same time. So it is a very ambitious project, and maybe it is not realistic in some parts. However, we need such visionary projects just as much as we need decisionmakers willing to adopt these ideas and help make them work.

New Dynamics.

We need, as I have said, a transatlantic as well as a global consensus on how to address climate change. And we need to renew transatlantic dialogue. There are many ways to achieve these goals. We can harmonize policy approaches as I have just suggested. We can also engage in more cooperative research. Here the energy and technology initiative currently promoted by Foreign Ministers Rice and Steinmeier may be a useful starting point.[17]

A first U.S.-EU high level meeting involving chief executive officers (CEOs) from energy, research, and venture capital took place in March 2007. Its goal was to hasten the innovation and deployment of new energy technologies across the Atlantic. Such initiatives need to accompany international negotiations carried out under the auspices of the UN. These negotiations might be triggered by dialogues between the G-8 countries and the five outreach countries, Brazil, India, China, Mexico, and South Africa. The aim of all these discussions is to develop a shared understanding of the nature of future commitments. Meetings like that in Heiligendamm in Germany could serve as springboards for a global approach on climate change and energy security.

In fact, international climate negotiations for a multitrack approach are scheduled to start in Bali at the end of the year. Lots of things will have to be agreed upon. Under discussion will be targets and timetables, technological partnerships and compensation mechanisms.

Conclusion.

In sum, despite the enormous challenges we still face, I am at this point somewhat optimistic. There are signs that a transatlantic consensus can be reached. If we build on our historical relationships and find ways to engage newly industrialized nations, the search to find a diplomatic solution to climate change may gain much needed momentum.

ENDNOTES - Tänzler

1. See Intergovernmental Panel on Climate Change, A Report of Working Group I (IPCC WG I), *Climate Change 2007: The Physical Science Basis – Summary for Policymakers*, World Meteorological Organization and United Nations Environment Programme, 2007; Nicholas Stern, "Stern Review: The Economics of Climate Change," Executive Summary, Cambridge, UK: Cambridge University Press, 2006.

2. See for a detailed analysis of both agreements, Sebastian Oberthür and Hermann E. Ott, *The Kyoto Protocol: International Climate Policy for the 21st Century*, Berlin, Germany: Springer, 1999.

3. See for an analysis, Chukwumerije Okereke, Philip Mann, Henny Osbahr, Benito Müller, and Johannes Ebeling, *Assessment of Key Negotiating Issues at Nairobi Climate COP/MOP and What It Means for the Future of the Climate Regime*, Working Paper 106, Oxford, UK: Tyndall Centre for Climate Change Research, 2007.

4. See for a more detailed discussion, Dennis Tänzler and Alexander Carius, "The Prospects for a Transatlantic Climate Policy," *Journal of Transatlantic Studies*, Vol. 2, No. 2, 2004, pp. 209-226.

5. See for recent GHG trends, The United Nations Framework Convention on Climate Change (UNFCCC), GHG Data 2006, Highlights from Greenhouse Gas (GHG) Emissions Data 1990-2004 for Annex II-Parties, Submitted under the United Nations Framework Convention on Climate Change (UNFCCC), 2006; available from *www.unfccc.int*.

6. See for a discussion of selected climate policy instruments, Sebastian Oberthür and Dennis Tänzler, "Climate Policy in the EU: International Regimes and Policy Diffusion," in Paul G. Harris, ed., *Europe and Global Climate Change: Politics, Foreign Policy and Regional Cooperation*, Celtenham, UK, and Northampton, MA: Edward Elgar, 2007, pp. 255-277.

7. Alexander Carius and Dennis Tänzler, "*Climat énergie: une nouvelle composante du contexte*" ("Climate and Energy: A

New Component in the Context of Security"), in *Les Cahiers de la Sécurité, Environnment, changement climatique et sécurité,* No. 63, 2006, pp. 157-186.

8. A binding target was not expected.

9. See Council of the EU 2007, Presidency Conclusions Brussels, Belgium, March 8-9, 2007, available from *www.consilium.europa.eu*.

10. See Dennis Tänzler, *The Diffusion of Renewable Portfolio Standards within the United States,* Paper presented at 3rd European Consortium for Political Research (ECPR), Budapest, Hungary, September 8-10, 2005.

11. The major lesson it learned was that more stringent allocation plans need to to be adopted for the first commitment period from 2008 to 2012. See for an analysis of the European Trading System, Damien Meadows, "The Emissions Allowance Trading Directive 2003/87/EC," in Jos Delbeke, ed., *The EU Greenhouse Gas Emissions Trading Scheme,* EU Energy Law, Vol. IV, 2006.

12. Emma Marris, "Western States Launch Carbon Scheme," *Nature,* Vol. 446, March 8, 2007, pp. 114-115.

13. Pew Center, August 2007 Chart, "Summary of Cap-and-Trade Climate Legislation Proposed in the 110th Congress," Washington, DC, 2007.

14. United States Climate Action Partnership (USCAP), "A Call for Action," 2007, available from *www.us-cap.org*.

15. See for a discussion, Daniel Bodansky, "International Sectoral Agreements in a Post-2012 Climate Framework," working paper prepared for the Pew Center on Global Climate Change, 2007.

16. See Gerhard Knies, "TREC: Renewables for Security and Stability," February 2007, available from *www.ecc-platform.org*.

17. See U.S. State Department, "Ensuring a Sound Energy Future," remarks with German Foreign Minister Frank-Walter Steinmeier and European Commissioner for External Relations and European Neighborhood Policy Benita Ferrero-Waldner, U.S.-EU Energy CEO Forum, Washington, DC, March 19, 2007.

Meeting the Challenge of Climate Change: The Role of the Corporation

Simon Rich
Louis Dreyfus Holding Corporation

The Challenge.

We are here today to discuss climate change. I have been asked to suggest what corporations can do to address the problem. I want to start by asking you to imagine a small circle surrounded by a larger circle. The larger circle represents the biosphere. As all the scientists in this room know, this biosphere operates under certain immutable laws. The smaller circle represents the global economy. This operates according to economic theories. I would suggest to you that we need to keep in mind that they are just that—theories. We are overshooting in so many areas. Our global economy is dumping too much waste in the biosphere for it to handle. We are overfishing. We are running out of oil. We are abusing our top soil. And our population is still growing quite rapidly. So the economy is pushing against the real limits of the biosphere to which we owe LIFE.

At the same time, we still operate according to the antiquated belief that there are no limits to growth. That theory worked well enough when the world was relatively empty. But can we afford to operate at the start of the 21st century the same way we did in earlier times? And what of the future? We have a stock market selling at 25 times earnings. This means that the investors are expecting these companies to achieve a significant rate of growth, and this is predicated within a mature economy. Second, China and India have rapidly developing economies and are using natural resources and creating waste at unprecedented amounts. Our biosphere is really not designed to handle this kind of exponential growth. If we are to expect two billion additional people over the next 30 or 40 years, maybe we should consider managing our global economies quite differently.

Our global economy is to a large extent a market driven economy. As we all recognize, accurate price signals are critical to the proper functioning of any market. But price signals can be distorted. Economists tend to assume that the substitutions of inputs to their economic models are without restriction. So, for example, if we are short of natural resources or labor, we can substitute technology. However, if we are getting the wrong price signals, we may not be able to make the substitution, or we may make one that is incorrect. Consider our current oil prices. At $60 a barrel, oil is cheap. It is certainly not properly priced because there are many unpriced attributes to the current price of oil. These include the cost of maintaining a military presence in the Middle East, the cost of health care as more and more people are affected by the poor air quality in most of our cities, and the one we are talking about today—climate change. Therefore, our $60 per barrel oil may really be costing us $100, and the resulting economic assumptions must be wrong.

We also find it difficult to do what needs to be done. To illustrate this point, let me tell you a brief and rather funny story. I am on the board of an organization called "Sustainable North Carolina." We give an award each year to the most sustainable business or operation. The year before last the award was won by the Marine Corps Air Station at Cherry Point, which is responsible for repairing airplanes. When the Marine

Corps colonel came up to receive his award, he made the following observations: He thanked everybody for the award. He said they would keep working at Cherry Point to defend democracy at home and abroad. Then he stopped and said, "But I'll tell you, I would never have won this award if I ran a democracy at Cherry Point. When I decided we were going to recycle on the base, believe me, we recycled. When I said we were going to cut water consumption, we did." He went through his whole list. His point is well-taken, and we all recognize the difficulty in moving a group without proper education or incentive. While I have been asked to focus on what corporations can do to meet the challenge, the story does illustrate an interesting point. The fact is that the Department of Defense (DoD) and the Army have a magnificent opportunity to lead the nation in sustainability—and, indeed, are doing so. Many corporations are taking the lead as well toward limiting green house gas emissions and moving toward a more sustainable business model in general.

Energy Efficiency.

So—to get to my central point—fossil fuel combustion leads to greenhouse gas emissions which lead to climate change. To deal with climate change, we obviously need to address fossil fuel combustion. What can we do? What are our opportunities?

Before I get to the opportunities, let us just look at the facts as I see them. Of all of our primary energy, 80 percent comes from fossil fuels. Two-thirds of the oil that still remains in the ground is in the Middle East. The CO^2 gas that results from the combustion of all of this fossil fuel is released into the atmosphere. Does it look to a rational person as if we have a sustainable situation here? Or does it look like we are headed for disaster? Our population obviously is increasing. So too is affluence. The demand of China, India, and Brazil for oil is increasing as their economies develop, and their need for energy services increase. If you use the metric of gross domestic product (GDP) divided by primary energy use, we have been doing a pretty good job in the developed world—energy efficiency has been increasing at 1 percent a year. The problem is that if world energy use is projected to increase at four times what it is today by the end of the century, we would have to increase energy efficiency 2 percent a year.

What I want to stress in my talk is the need for energy efficiency. This is where opportunities lie. We are facing some very serious issues. They are challenging all of our scientists and our engineers. They are made more complicated by our sociopolitical and economic structures and even our democracy. We have very disparate views on how to deal with the problem, and the disagreements are fiercely held and slowing us down. And we have to deal with the difficulty of how to convert the developed world to sustainable practices, while at the same time the developing world wants to achieve a certain standard of living. And we have to deal with human needs as opposed to the need for economic growth. Finally, we just cannot remain completely dependent on fossil fuels. We have to figure out a whole new trajectory.

We also face some technical questions. We have to know how and if CO^2 can be sequestered. Can we really capture CO^2? Can we put it in the ground? Can we pipe it and use it in industrial processes? Are there cost-effective ways to use CO^2? That means research. It means more science dollars. The same is true of nuclear energy. Can we store the waste safely? When the President of Greenpeace comes out and says, "I'll support a nuclear plant that's been properly thought out," when the environmental community is ready to embrace nuclear power to avert CO^2 emissions, you know that there has been a major sea change. Bio-fuels are being touted as an answer all over this country, including in North Carolina. However, lingering questions remain. When you use bio-fuels, you are placing energy demands against the food supply. The current model for agriculture in the United States and the developed world is basically an industrial monoculture, facing very serious problems from the standpoint of water use, soil erosion, and fossil fuel use.[1] Finally, there are questions about solar power. Can we make solar power affordable?

In the limited time that remains to me, let me suggest some of the ways we could become more efficient in the area of electric power and transportation, both key consumers of fossil fuels and about equally responsible for the emission of greenhouse gases.

In the United States, 70 percent of all electric power comes from fossil fuels—about 50 percent comes from coal, and nearly 20 percent from natural gas and petroleum. We have had a major nuclear program here, which is stalled, and we are basically doing nothing on renewables. So we are totally dependent on fossil fuels. Now this is a bit of a disaster, but also it could become an opportunity for us, if we dedicate ourselves to becoming more energy efficient.

If you look at a map of the United States, you can see that coal burning is fairly evenly distributed until you get to the western systems power pool. California buys a lot of electricity from Montana and Arizona. So we Americans are totally coal-dependent, and we are emitting greenhouse gases all over. North Carolina is downwind of a major coal-burning region. Our Blue Ridge Mountains are being devastated right now from the emissions coming out of the east north central region of the United States.

Our current technology is very inefficient today. When we put 100 units of coal, or anything else, into our system, we are getting nine-and-a-half units of energy service coming out. This inefficiency presents a tremendous opportunity, but one which we can only take advantage of if we spend a significant amount of money on research and science. If you want to involve members of the business community, you have to provide them with incentives to invest private dollars in the area of energy efficiency. In addition, we need intelligent public policy or we need the military to take the lead.

Clearly, if you can cut the number of units of energy that you need by becoming more efficient in your energy use, you have a tremendous leverage on the amount of fossil fuels you are using. It means that small amounts of energy efficiency translate into large amounts of primary energy saved. This will help us as we institute significant energy efficiency campaigns.

Here in North Carolina the utilities commission recently turned down Duke University's application to build two coal-fired power plants down near Charlotte, two 800 megawatt power plants. The Commission told Duke it could build one—which they probably cannot do since their economic calculations were based on their building two. It also told Duke Power, "And you have to spend $50 million a year on energy efficiency in the state for the next 10 years." That money spent on energy efficiency is going to have a big effect in North Carolina, and the challenge to any regulated state like

North Carolina is for the Utility Commission to figure out a way—and they will—to incentivize the utility to spend that money on energy efficiency to create the NEGAWATT, that is with an "N," and earn a return on the megawatt rather than the megawatt which is generating capacity they must build.[2]

The utilities commission had a study done for them by a very reputable consultant where a new kilowatt of electricity might cost $2,000 or more to construct, and that same amount of electricity can be saved through greater efficiency and thus reduce the need, the demand, at one-third the cost. When you institute an energy efficiency campaign, it is independent businesses that get involved. You need people to do the heating, ventilating, and air conditioning (HVAC); the hot water heating; and selling the Energy Star appliances. It is not massive business but local business that begins the process. So the dollars stay in the place where energy efficiency is being implemented. So it is very positive for local economies, and it creates jobs. The study suggested that 22,000 jobs would be created in North Carolina with the institution of this energy efficiency campaign. It is a million jobs around the United States.

This is the shocker for me—you all are probably aware of this—but if we put one compact fluorescent bulb in each house in the United States, we would save two coal plants or take 1.3 million cars off the road. The savings are out there, but we are not doing it. We do not have the political will to do it. So who are the leaders? Australia: they said, "No incandescent lights." The European Union: they said, "No incandescent lights." North Carolina has got a bill: "No incandescent lights." And look at Fort Bragg's web site on sustainability: It is phenomenal what Fort Bragg is suggesting they are going to do.

You are probably very familiar with the arguments in favor of using renewable energy. This is indeed important. If we instituted a 20 percent renewable energy and efficiency portfolio standard in the United States, we could save $76 billion a year. If 50 percent of that was from efficiency, that money would go into local economies throughout the United States. People in the United States would not be looking for jobs at Wal-Mart any more, believe me.

And we haven't even begun to price the externalities. The Clean Smokestacks Bill was passed in North Carolina 3 years ago simply because we drew attention to what it cost to treat asthma in children and other health problems caused by the dirt we were putting out of the smokestacks. It was passed by the legislature because they realized that spending $2 billion to clean up the smoke stacks was better than spending $8 billion in health care in North Carolina. That is just one of the externalities that has not been priced.

My last point relates to transportation. You know, we are rejecting Corporate Average Fuel Economy (CAFE) standards, and here is what happened. We got serious about CAFE standards in 1980. We stated that mileage per gallon had to increase, and look what happened—our petroleum consumption went way down. But we relaxed the standards because the auto industry lobbied millions of dollars every year, and they kept it from happening. Rick Wagoner, the Chief Executive Officer of General Motors, stood up the other day and said, "We don't need CAFE standards." That is the mentality that is hard to get past. Anyway, we have the same ability to be efficient when it comes to the automobile that we do when it comes to the provision of electricity. Of the energy that goes into an automobile, 87 percent is wasted. We can easily reduce this waste by increasing CAFE standards, saving tremendous amounts of money each year. If we can get to 43 miles to the gallon—we've got the technology to get to 80—we would save three million barrels a day.

As to bio-fuels—they have a place. Bio-mass has the ability to produce a whole range of fuels. The feedstocks range from wood and any woody vegetation to specific energy crops and agricultural waste. We are using the worst one possible—corn. As you saw in the *Wall Street Journal* today, corn planting has hit the highest level since 1944. Corn is commanding over $4.00 a bushel. Farmers are making a fortune. Well, the reason corn is worth this much is because it reflects a supply curve for ethanol produced from corn. It is uneconomic without the subsidy. You put the subsidy in, and everybody is growing corn. We have to move from corn to cellulose, which is cellulosic ethanol process, and we will have a reasonable bio-fuel from an energy balance and environmental perspective. Again: research, development, science. It is possible, we are just not doing it.

Bio-fuels are a substitute for up to 2.5 mm barrels of oil each day. That is more than comes out of the Persian Gulf or the Arctic National Refuge, which is just hardly a sneeze compared to the amount of oil we use. Even though it is a fraction of our use, it amounts to $55 billion a year.

What we need is what I call an authorizing focus. That is to say we need public policy to set out the rules for business. If we leave the free market to its own devices, what happens? It operates according to a "who's the fastest gun?" principle and it does not price things or allocate capital properly. The free market is great at pricing things in the short-term, but we need to price the externalities, and business needs to be guided by informed public policy.

We also need leaders. There are some. (If you want to read a fascinating story, read about Goldman's intervention in the TXU deal.) But there are also laggards. The biggest laggards are the U.S. policymakers and the White House. They have been stonewalling this whole notion of a sustainable energy policy bowing to the demands of special interest groups and the money they dispense. It is a travesty—General Motors, stonewalling; Exxon, stonewalling; the Southern Company, stonewalling; and TXU was until they got bought out, and now an enlightened ownership and leadership there are going to make a difference. Chrysler has the worst fuel efficiency of all the auto marketers, and their cars are becoming dinosaurs.

So to sum up: We need to follow a new paradigm. Below are four points which can help business become more efficient and contribute towards the solution of climate change.

1. We need to put economics in a broader context. As it stands, we define things too narrowly. We do not price externalities.

2. We need to lengthen our accounting periods. I used to run a public company; you live quarter to quarter. You cannot think that way in a world like we are facing today.

3. We have to think in a less linear fashion and think instead in circles. We need to work out how we can recycle things. At the moment, we put something in at one end, use a little bit of it, and dump 94 percent of it out at the other end.

4. Finally, we need an authorizing focus. Government has got to stand up and be government, and lay down the rules for a sane future for all of us.

ENDNOTES - Rich

1. On a slightly different note, it is worth remarking that modern agriculture poses yet other problems. The South American farmer is using the same seeds and the same genetic stock as the guy in Iowa or the guy in North Carolina or the guy in Europe. So our genetic stock is getting narrower and narrower and narrower and narrower, and we are opening ourselves up for a major genetic failure in our food supply.

2. Amery Lovins (founder of the Rocky Mountain Institute and a recognized authority on energy efficiency) coined the term "negawatt," and I stole it because I liked it.

Climate Change in the American Mind

Anthony Leiserowitz
Yale University

Introduction.

This presentation is entitled "Climate Change in the American Mind." We will consider how Americans perceive the risks of climate change, the severity and the likelihood of various impacts, and what kind of policies they support or oppose. We will also, however, take a look at some of the underlying reasons why there are such strong and sometimes polarized opinions about this issue. Why is it that some people believe climate change is a very serious problem, while others do not? Why do some people support aggressive climate policy, while other people do not? And finally, we will identify several distinct "interpretive communities" within the American public—different audiences each of which perceives the risks of climate change and other hazards (e.g., terrorism, nuclear power, pesticides, marijuana use, etc.) in a uniquely patterned way.[1]

To begin, let us step back to consider the broader context. It bears repeating that, with only 5 percent of the world's population, the United States alone produces about 25 percent of the world's greenhouse gases. Thus the American people are critical to the ultimate solution of this problem through their individual consumer behavior and energy use, their political preferences, the leaders they vote for, and the kinds of climate change policies that they will support. With regard to the many potential threats climate change poses to national security, it is important to ask what kinds of military actions the public will support.

Will they support the use of either soft or hard military power to address climate change impacts around the world? Will they support the deployment of American forces to deal with potential climate-induced crises ranging from famine relief to intervention in resource conflicts or emerging pandemics around the world? We do not yet have answers to these questions regarding the military, but we are beginning to understand how the public perceives the risks of climate change.

Global Warming and the American Public.

To begin, it is important to recognize that the public as a whole is now very aware of global warming. About 95 percent of Americans have heard of global warming and have at least some rudimentary knowledge of it. Further, a 2006 Pew study found that about 41 percent of Americans said that it is a very serious threat, while another 32 percent said it is somewhat serious, while only about a quarter said that it is not serious or not a problem. So most Americans are clearly aware of global warming and think it is a serious problem.

In 2003, I conducted an in-depth study of American opinion on global warming and found strong bipartisan support for a number of national and international policies. Americans strongly supported the Kyoto Protocol, and strongly believed that the United States should act to reduce its emissions regardless of what other countries do. They strongly supported higher fuel economy standards, regulating carbon dioxide as a pollutant, and shifting subsidies away from the fossil fuel industry towards the renewable energy industry. There was very strong bipartisan support for all of those policies, but interestingly, very strong bipartisan opposition to carbon taxes. Americans were strongly opposed to paying $300 more a year as a household as part of a business energy tax and strongly opposed to a 60 cent per gallon gasoline tax.

So, again, the American public are aware, they are concerned, they want action at the international or national level, but do not support higher energy prices in the form of carbon taxes. That is the current political reality.

In 2004, in another nationally representative survey, I asked Americans which issues they thought should be the top priority for Congress and the President. I found—and this is consistent with many other surveys—that global warming was a relatively low priority, just as the environment as a broader issue is almost always at the bottom of these kinds of priority rankings. Global warming was well below terrorism, the economy, healthcare, education, the budget deficit, etc. Today we would see the Iraq War, of course, as a leading national priority as well.

So global warming was a relatively low national priority, but what about within the environmental category? How did global warming compare to other environmental issues? It has risen in recent years, but it was still well below water pollution, air pollution, toxic waste, and even the ozone hole. The relatively low priority of global warming helps us understand why there has not been more action to date. There has been a lack of political will in the United States. So why is that?

Understanding the American Response to Global Warming. In part, that has to do with the way Americans perceive the risks of global warming. My research demonstrates that across the board, the public generally perceives climate change as a moderate risk that will primarily impact people, places, and ecosystems distant in time and space. They believe that global warming is a greater threat to nonhuman nature than to human beings. They believe that water shortages, increased disease rates, and lower living standards, are only moderately likely. Importantly, they believe that each of these impacts is more likely to occur globally than at the local level. Americans tend to think of climate change as a distant problem, something that is going to affect other people far away—small island countries, poor people in the tropics, etc.—not Americans—and distant in time—not for another 50 to 100 years, if ever. Thus it is not a particularly salient issue to most people.

Another way to examine this is to take a deeper look at the connotative meaning of global warming—the thoughts, feelings, and images that are evoked by the term. To assess these deeper meanings, we asked respondents in a representative survey a seemingly simple question: What is the first thought or image that comes to your mind when you hear the words "global warming?" These free associations were then compiled and content analyzed to identify the meanings of climate change in the American mind.

First and foremost, Americans most frequently associated global warming with melting ice. Americans mentioned the loss of sea ice in the Arctic Ocean, ice shelves breaking off in the Antarctic, and glaciers melting around the world; and they rated these as very negative things, as very bad things. But most Americans do not live in the Arctic, very few Americans live in the Antarctic, and most of them do not live anywhere near a glacier. While the melting of ice around the world is considered a bad thing, it is not a direct threat to the lives of most Americans.

The second most frequent set of associations were to heat and warming temperatures, e.g., things will get hotter, temperatures are rising, etc. Third, many Americans thought of the impacts of global warming on nonhuman nature, including other species and ecosystems, not people. Fourth, many Americans associated global warming with the ozone hole. This and other research has demonstrated that many people in the United States and around the world continue to confuse and conflate global warming with the ozone hole, with many people thinking they are either the same thing, or that the hole in the ozone layer is the cause of global warming.

The fifth largest set of associations came from people I will call "alarmists," whom I will discuss further in a moment. Sixth, were a set of associations to flooding and sea level rise; these tended to be a little more concrete, with people mentioning actual places, like Manhattan going under water. Next came references to climate change or to changing seasons, neither of which were rated as particularly negative or bad. Finally, the eighth largest set of associations came from people I will call "naysayers" who did not see global warming as a problem at all. Thus, we found that the top four sets of associations to global warming, representing over 60 percent of all responses, were to melting ice, generalized warming trends, and the impacts on nonhuman nature—none of which represent direct threats to the lives of most Americans—or to a completely different environmental problem. These predominant connotative meanings of global warming help us understand why Americans have perceived this issue as a moderate threat and a relatively low national or environmental priority.

Missing Links. One of the most important findings was what we did not find. We did not find any Americans who made the connection between climate change and human health. There were no associations to heat stroke, asthma, or infectious diseases. Likewise, there were no associations to extreme weather events. People did not associate global warming with hurricanes, tornadoes, or drought. Now this survey was conducted prior to Hurricane Katrina. In the aftermath, some in the scientific community and the media indirectly linked more powerful and intense hurricanes to global warming. So one might think that hurricanes are now firmly linked in the public mind to global warming. I have conducted several other similar national studies since Hurricane Katrina, however, and have found only a few scattered associations to hurricanes. When directly asked, however, more people say that global warming is making hurricanes worse than said this before Hurricane Katrina hit. But when not primed by the question itself, hurricanes currently rarely come to mind when people think about global warming. Finally, we found no associations to the kind of national security implications that we have been talking about here at this conference; most Americans are not connecting the dots between global warming and national security.

Interpretive Communities of Risk. So now let us turn back to the two groups I mentioned before, alarmists and naysayers. Here I would like to introduce the concept of interpretive communities of risk. We often speak of "the American public," but this is misleading. The United States is actually a set of diverse publics—for example, there are very different groups within the United States that respond to risks in very different ways. They may confront the same issue, but they often approach it from completely different perspectives and come away with very different conclusions. Global warming alarmists and naysayers are two examples. "Alarmists" refers to respondents who associated global warming with catastrophic and even apocalyptic impacts. They often imagined impacts well beyond the worst-case scenarios projected by the scientific community. For example: "like after nuclear war," "it is going to kill the world," or "death of the planet."

By contrast, some Americans were "naysayers"—respondents who denied the reality or seriousness of global warming. Naysayers offered six distinct reasons why they believed climate change is not a problem. Some flatly denied it—it is just not happening. Others said it is happening, but it is natural, not human-caused. Others argued it might be happening, but did not think the science had been proven yet. Some doubted based on their own recent personal experience, e.g., "It was -10 here last night, it can't be warming." Some blamed it on media hype. Finally were the conspiracy theorists, who said things like "a hoax," "a fraud," or "scientists making up statistics for their job security." For conspiracy theorists, the accumulation of ever more scientific evidence probably only reinforces their sense of conspiracy, thus it is very hard to convince these particular people.

So how did these two groups compare with each other and the rest of the public? I compared each group's average risk perceptions across 13 different measures, including the seriousness of current impacts, the threat to nonhuman nature, the likelihood of local and worldwide water shortages, infectious diseases, diminished standards of living, etc. What I found was that, not surprisingly, naysayers rated all of these items as a low risk, whereas alarmists saw them as a high risk. What was interesting, however, was how the other 75-80 percent of the public rated these items. The rest of the public was much more similar to alarmists than they were to naysayers. Naysayers, it turned out, were very different from the rest of American society. When I compared the average support of these three groups for a variety of climate change policies including support for the Kyoto Protocol, increasing fuel economy standards for cars and trucks, regulating carbon dioxide as a pollutant, shifting government subsidies from the fossil fuel industry to the renewable energy industry, a carbon tax on business energy use, or a carbon tax on gasoline; again there were very large differences between alarmists and naysayers. However, the rest of the public's support for these policies was much closer to the level of alarmists than naysayers.

One other key finding of this study was the critical role values play in mediating the way people perceive global warming. I found a strong positive correlation between egalitarian values and climate change risk perception. In other words, the more egalitarian their values, the more concerned respondents were about global warming. The less egalitarian, the less concerned. For example, naysayers strongly disagreed with a variety of egalitarian statements. They did not support affirmative action, they did not think the world needs a more equal distribution of wealth, they did not support more participatory decisionmaking, or government efforts to end poverty, and so on. By contrast, alarmists strongly supported these statements.

On the other hand, naysayers held strong individualistic values. For example they were much more likely to agree that, "The government should just get out of our way and leave us alone." "Life sorts out those that try hard from those that do not." "Government has no right to regulate personal risks." And so on.

So who were the alarmists and naysayers? What were their socio-demographic profiles? Alarmists tended to be slightly more liberal and slightly more Democratic. Naysayers, on the other hand, were a very distinct group. They were overwhelmingly white, male, conservative, Republican, highly religious, and often got their news and information from talk radio.

Another important point is the critical role of trust in how people respond to risk. Trust has been called the currency of the realm and is absolutely central to society's ability to deal with an incredibly complicated, often abstract, and unfolding problem like climate change. Trust is important because climate change is difficult to understand, most people know little about the causal mechanisms of climate change or the most effective solutions, and it is just one of a myriad of other risks that people are now forced to confront with limited personal expertise or experience. Furthermore, most people have very busy lives, with many other issues and personal affairs competing for their limited time and attention. As a result, many people look to their opinion leaders, to people they trust to guide them through unknown and uncertain terrain—people they trust to tell them the truth.

As part of this study, I asked respondents, "How much do you trust each of the following groups to tell you the truth about global warming?" I found that naysayers were again very different than both alarmists and the rest of the public. Naysayers strongly distrusted the media. They also strongly distrusted environmental organizations. As a result, environmentalists are perhaps the least effective communicators about climate change to this particular audience. Naysayers were also less likely to trust scientists and doctors, although they trusted these sources more than most others. Surprisingly, naysayers were even less likely than other Americans to trust their own friends and family to tell them the truth about global warming.

As you can see, these two different interpretive communities—alarmists and naysayers—each approached the issue of global warming from very different perspectives and drew very different conclusions about it as a threat, or how society ought to respond. After this initial study, my colleagues and I became interested in exploring this idea of "interpretive communities of risk" beyond just climate change, so we did another nationally representative survey in June 2005. In this study, we asked, "How great a risk are each of the following to American society?" Items included a variety of national security, technological, environmental, health, and moral "hazards," including terrorism, the Iraq war, global warming, nuclear power, pesticides, genetically modified food, gun control, marijuana, legal abortion, and homosexuality. We then conducted a segmentation analysis to look for groups with different response patterns and identified five distinct groups, or "interpretive communities of risk" within the American public.

The first is what we labeled "alarmists" (12 percent)—a group that rated all of these hazards as high to very high risks. By contrast, another group, labeled "optimists" (21 percent) rated all these hazards as a relatively low risk. We then identified two groups that were almost the mirror opposite of one another. The first we labeled "the religious right" (16 percent) and they rated items like abortion and homosexuality as a very high risk. Opposing these views was a group we labeled "the liberal left" (14 percent) who rated those same items as a very low or non-existent risk. By contrast, the liberal left rated things like global warming and the Iraq war as a high risk, whereas the religious right rated them as a relatively low risk. Finally was a group that we labeled the "mainstream." This was the largest of the five groups (37 percent), and had a relatively moderate level of risk perception across all the items.

Now let's return to the specific issue of global warming. This study found that over 60 percent of the American public fell into three interpretive communities of risk—alarmists, liberal left, and mainstream—all of whom already saw climate change as a high to very high risk. The primary opposition on this issue, however, came from the other two groups—the religious right and optimists. We now believe that the group of climate change naysayers I discussed earlier is actually comprised of these two groups.

Interestingly, the religious right and optimists share a number of characteristics in common. Both are predominantly white, male, conservative, and Republican. Optimists, however, tended to have significantly higher incomes than the religious right. The two groups also differed greatly in their religious orientations. Unsurprisingly, the religious right was very, very religious. Optimists, however, were not. This difference helps to explain why these two groups strongly disagreed with one another about the threats abortion and homosexuality pose to American society, with the religious right respondents perceiving these issues as great risks, while optimists perceived them as very low risks.

Now back to global warming—which both the religious right and optimists rated as a relatively low risk. Each of these groups is currently undergoing a significant internal transformation in the United States. Within the religious right, a large number of evangelical leaders have recently broken with their peers to argue that global warming is indeed happening, that humans are at least partly responsible, and that this is a moral issue that Christians are called to confront. These leaders justify this new position by arguing that in the book of Genesis, God commanded human beings to till and tend his garden, and that the environment is part of our stewardship responsibilities on the earth, to care for God's creation. Thus global warming is a moral imperative. Secondly, many argue that action on global warming flows directly from their long-standing missions to help the poor and needy, such as famine and poverty relief around the world. To paraphrase, "How can Christians devoted to these acts of mercy in good conscience ignore a problem that is going to push millions of people into the same kind of circumstances that we are there to help them with?" Importantly, these are arguments that resonate within the religious right's own strongly-held value system. Yet these specifically Christian arguments may not resonate with other audiences. There are, however, many roads to Damascus. Different people, starting from very different moral and ethical standpoints, can at times reach the same conclusions and work together in common action, albeit sometimes for different reasons.

By contrast, optimists tend to be more libertarian and entrepreneurial in orientation and may represent the economically conservative wing of the Republican party. Interestingly, we are currently seeing a tremendous shift within the business community on climate change. Companies are moving quickly to address this issue both as a response to the physical and legal risks of climate change and to the enormous market opportunities for those who develop the solutions. Solving the climate change problem is ultimately about moving the world economy from its current foundations in fossil fuel energy to noncarbon emitting energy sources. Reducing global emissions 60 to 90 percent below 1990 levels is going to be a tremendous challenge and will involve hundreds of billions, if not trillions of dollars. Somebody is going to make that money. Further, this transition will produce great technological innovations and the creation of whole new jobs and industries. There are now a lot of companies looking at this emerging energy future and moving aggressively to exploit this new market.

Over the next few years these internal transformations are likely to reshape each of these groups, the religious right and optimists, and the elites of both groups and their like-minded constituents are going to increasingly perceive global warming as a significant risk and support serious action to deal with it. Of course, some members within each of these groups will never accept the reality, human causation, or importance of this issue, but they will likely become a smaller and more marginal minority.

Communication Implications.

So what are some implications of this research for climate change communicators? Well, first of all, communicators need to highlight the potential local and regional climate change impacts for Americans. Tip O'Neill, the former Speaker of the House of Representatives, once said, "All politics is local," and to the extent that is true, people need to understand that, yes, global warming will impact people and places far away; yes, it will impact polar bears and other species; but it also has real consequences for Americans. Second, climate change is not a temporally distant phenomenon—it is happening right now, and there are a lot of examples around the world. As just one example, I have just completed a 2-year project in Alaska, and, as the Arctic Climate Impact Assessment has reported, the impacts there and across the Arctic are dramatic—sea ice disappearing, permafrost melting, record forest fires and insect pests, changes in species distributions and migrations, infrastructure crumbling, coastlines eroding, and forced relocations of entire communities. Third, it is important to highlight the potential impacts of climate change on human health, on extreme weather events, and, especially for this conference, the potential impacts on our national security. Most Americans have not yet connected these dots. Finally, it is critical that we recognize that the "American public" is in fact a set of multiple and diverse audiences. Effective climate change communications will often require tailoring the message and the messengers for these specific audiences.

ENDNOTES - Leiserowitz

1. For further details, see A. Leiserowitz, "American Risk Perceptions: Is Climate Change Dangerous?" *Risk Analysis*, Vol. 25, No. 6, 2005, pp. 1433-1442; *idem.*, "Climate Change Risk Perception and Policy Preferences: The Role of Affect, Imagery, and Values," *Climatic Change*, Vol. 77, 2006, pp. 45-72; *idem.*, "Communicating the Risks of Global Warming: American Risk Perceptions, Affective Images and Interpretive Communities," in S. Moser and L. Dilling, eds., *Creating a Climate for Change: Communicating Climate Change and Facilitating Social Change*, Cambridge, UK: Cambridge University Press, 2007, pp. 44-63.

Commentator

Marvin S. Soroos
North Carolina State University

As a political scientist, I have been dealing with global environmental issues and international responses to them throughout most of my career. I was one of the first international relations specialists to teach and conduct research in the field of global environmental politics, law, and policy. I have been teaching courses in this field since the mid-1970s.

I would like to draw attention to a couple of general themes that have emerged in our proceedings thus far. The first is the dynamic nature of the science of global climate change. A few years ago, several of us here participated in a workshop at the Sandia National Laboratories on the topic of "abrupt global climate change" and its implications for national security. In preparation for the conference, the participants were asked to read an article by Spencer Weart which highlighted how, over the past several decades, there has been a paradigm shift in the assumptions of climate researchers on how rapidly significant climate changes could take place.[1] As recently as 50 years ago, it was generally believed that major changes in the global climate, including those that might be induced by human activities, could occur only very gradually over thousands of years. By the 1960s and 1970s, further research suggested that severe climate change could occur in a matter of centuries, and in the next decade the time frame was further adjusted to a century or less. By the mid-1990s, research on ice cores extracted from the Greenland ice sheet revealed that over the past 100,000 years, there have been several episodes of "abrupt" climate changes occurring in as little as a decade. These naturally occurring rapid changes raise the possibility that human activities might push the climate system past tipping points that could rather quickly usher in a very different climate regime. In the coming decades, we must be alert to the possibility of other major adjustments in our basic scientific understanding of the dynamics of global climates and human influences on them.

Furthermore, as I review articles on internet news-feeds on climate change, I have been impressed by how many scientific research reports are being released these days on the diverse impacts of global climate change. The momentum of these reports seems to have accelerated to the point that sometimes I wish we could put a hold on the science of climate change impacts for 5 years to allow us to catch up on these developments, to reflect on their many implications, and to decide how to adapt to them before there is another cascade of scientific findings. But unfortunately, that is not going happen. The dynamic nature of the science of climate change and its impacts makes this a very challenging field to follow.

The other general observation I would make—and this leads into the presentation by Dennis Tänzler—is how frustrating it is to get nations to work constructively together to create international and global environmental regimes. We have had some success stories, such as the international response to the problem of depletion of the stratospheric ozone layer. At least for now it looks like the ozone regime is a success story. However, when it comes to the global climate change issue, the international community has scarcely moved beyond square one. From a national security perspective, it is imperative that we think about how the United States can become a more constructive partner in international efforts to address climate change than it has been in the past.

Most Americans do not understand much about international law. They do not grasp what it is to sign and ratify treaties or the obligations that go with them. It is encouraging, however, that polls suggest a majority of Americans think the United States should ratify the Kyoto Protocol, but most respondents probably are unaware what is in that treaty and what it would require of them. I wish I could be more optimistic than Tänzler about the prospects for really significant cooperation in addressing climate change between Europe, the United States, and other parts of the world. It is encouraging that talks are taking place. However, in international meetings, such as the 12th Conference of the Parties to the United Nations (UN) Framework Convention on Climate Change in Nairobi in 2006 and the recent Group of Eight environmental ministers' conference, the United States remained adamant that it will not agree to mandatory limits on greenhouse gas emissions, which most of the rest of the developed world agree are needed and are committed to under the terms of the Kyoto Protocol.

The Kyoto Protocol at best can be only a small first step toward addressing the problem of global climate change. If all of the developed countries, including the United States, were to achieve the emission limits that they agreed to in the protocol, their combined greenhouse gas emissions for the period 2008-12 would only be about 5 percent lower than in 1990. Such reductions would be more than compensated for by massive increases in greenhouse gas emissions by the developing countries, in particular China and India. To make significant progress toward dealing with this problem, developed and less developed countries will have to go much further than they have so far. Unfortunately, diplomacy and negotiations are complicated by gross inequalities in per capita emission. Americans emissions per capita are generally about twice those of Europeans, and yet it is the United States that rejected the Kyoto Protocol and continues to refuse to make any commitment to capping its emissions. Developing countries have good reason to refuse limits on their greenhouse gas emissions as long as their per capita emissions are a small fraction of those of the highly developed countries.

Let me turn to the second presentation by Simon Rich. I applaud many of his suggestions. It is just wonderful to hear those all put together, offering hope that, if we could get the politics right on this and get government policy to back these initiatives, we could deal with this problem of climate change.

As Rich suggests, growth is so much a part of our economic paradigm these days that environmentalists are fighting a very hard battle to try to make much of a change in it. Recently, in an article in the March/April 2007 issue of "Mother Jones," Bill McKibben argued that growth was not always the predominant economic value; in fact, the emphasis on growth dates back only 60 or 70 years. Much of the article discusses how little economic growth has enhanced human happiness, particularly in the developed world. Up to about $10,000 gross national product per capita, economic growth does bring about substantial increases in human happiness, but beyond that additional income adds very little to satisfaction with life. Americans are experiencing economic growth, consuming more, and emitting more pollutants with little increase in their happiness. Altering this preoccupation with growth is going to be a challenge both in the United States and elsewhere. For example, Vaclav Klaus, President of the Czech Republic (and an economist by profession) has argued recently that environmentalists are a greater threat to the open and free society than communism was. That is a pretty strong and disturbing statement.

I also share Rich's concerns about bio-fuels, and in particular those based on corn. We need to consider the potential unintended consequences of a wholesale rush to bio-fuel development, such as impacts on the production and price of food, degradation of agricultural land, loss of forests, and pollution from the application of pesticides and fertilizers. Furthermore, what quantities of fossil fuels will be required to produce each unit of energy from bio-fuels?

Anthony Leiserowitz offered many fascinating findings from his opinion research on the differences between climate change naysayers and the alarmists. As he spoke, I thought back to a book that I read as an undergraduate by E. E. Schattschneider entitled *A Semisovereign People: A Realist's View of Democracy in America.*[2] I specifically recall Schattschneider's argument that intense minorities in the United States almost always win out over apathetic majorities in the American political process. I have kept this little nub of theory in my mind ever since. This insight seems to holds true for the politics of climate change.

I also agree that Americans have a phobia about taxes. The last thing they want governments to do is increase them. I wonder if we could persuade them to pay more gasoline and carbon taxes, while at the same time reducing income taxes an equivalent amount. That may be too complicated an idea for politicians to sell to the American public.

It also appears that Americans tend to demand action when they believe their health will be adversely affected. Thus, perhaps we need to do more to increase public understanding of the implications that climate change is having for human health. There are indications that the severity and urgency of the global climate change problem is more widely appreciated by the American public. Leiserowitz's remarks about the evangelical community are encouraging. I have also noticed such a trend as some leaders of the movement are calling upon Christians to practice stewardship toward preserving the planet's natural system while warning of the seriousness of the consequences of global climate change. Such a change in the perspectives of the evangelicals would be significant both because of their numbers and their suspicion of the values of environmentalists.

Altering basic attitudes does not come easily. Psychological research suggests that people seek out those information sources that reinforce their views and avoid those that challenge their beliefs. Those who are skeptical of climate change are going to continue listening to Rush Limbaugh. It may take another one or two major catastrophic types of events—heaven help us—to really shake up the thinking public's opinion on the severity of the climate change threat to the point that the American public will support the decisive actions that will be needed to address the problem.

Discussion

Q: Leiserowitz congratulated the business community on paying attention to the issue of climate change, but Rich asserted that the business community needs incentives. Where is the disconnect?

Rich: We do need an authorizing force to set rules and make sure that everyone plays by the same rules. We also do need incentives in some areas. The problems of how to dispose of nuclear waste or create cellulosic ethanol will not be solved short of money and research. The private community cannot afford the risk, so incentives or a public-private partnership is needed. In other areas—take energy efficiency, for example—the benefits of taking action are obvious, and people are moving ahead.

Q: What percentage of the population were the naysayers and how many people were surveyed?

Leiserowitz: I have identified them using several different methods and with different surveys over the years. About 12 percent are hard-core naysayers, and perhaps another 15 percent are still skeptical. The numbers of respondents in these surveys have ranged from 673 to over 1,000.

Q: Has American intransigence over climate change led to a weakening of the U.S.-European Alliance?

Leiserowitz: There is not much data on global attitudes toward climate change. But Europeans in general are much more concerned about this issue, much more supportive of policy, and much more supportive of taking individual action on the issue, than Americans. Some are quite angry. Many in England and other European countries care a great deal about this issue, and they do not like our particular administration's policies or approach to it.

Tänzler: European opinion is regularly polled. Polls show that Europeans are very concerned about this issue and suggest that more people in Europe than America would be willing to accept more taxes to deal with it. European response to the United States is somewhat surprising though. At the beginning of the century, in 2001 and 2002, there was really a kind of Bush-bashing. But perceptions are changing. People are coming to recognize that the administration does not speak for all Americans. You do find many articles in German and British newspapers talking about changing perceptions in America. They do not just point to Gore and his Oscar. So I am quite optimistic.

Q: The United States and the EU have argued over the wisdom of using genetically modified grains that are drought-resistant in drought-afflicted areas, especially sub-Saharan Africa. Has any compromise been reached?

Tänzler: As far as I know, the EU is very reluctant. When it comes to sub-Saharan Africa, this is in part because other areas have greater priority like transboundary water cooperation or even the expansion of renewable energies.

Leiserowitz: The international survey data on attitudes toward genetically modified food shows that the strongest support for biotechnology—stronger even than the United States—is to be found in the developing world. Americans tend to be technological optimists—to believe that new technology can solve problems. By contrast, Europeans are much more skeptical about biotechnology and genetically engineered food. There are a whole host of reasons for that, including an important cultural dynamic—food is very important culturally in Europe. But it is striking that many developing countries are eager to get these technologies, for good or for ill.

ENDNOTES - Discussion

1. The article had appeared in *Physics Today*, June 11, 2004.

2. E. E. Schattschneider, *A Semisovereign People: A Realist's View of Democracy in America*, Austin, TX: Holt, Rinehart and Winston, 1960.

CHAPTER 5

PREPARING FOR A CHANGED ENVIRONMENT

The authors here consider, in different ways, what should be done to prepare for the kind of chaotic and uncertain international environment that might come about in the event of global climate change. Dr. Michael Glantz discusses the importance of developing early warning systems that work. Dr. Robert Dorff stresses the importance of promoting good governance across the world. Dr. Henry Gaffney introduces the reader to a variety of different scenarios. He also suggests what steps we might take in the face of these different eventualities.

Achillies' Other Heel:
Early Warning Systems

Michael H. Glantz
Center for Capacity Building,
National Center for Atmospheric Research

Introduction.

My purpose today is to talk about early warning systems. I like to think about early warning systems as Achilles' other heel. To be sure, the hero was vulnerable where Thetis grabbed on to his one heel, but he did, after all, have another, and it was never tested. We need to remember that—and see if we cannot find something that works when it comes to dealing with the problem of climate change.

I was educated as a political scientist, but now I consider myself to be a social scientist. I started out studying foreign involvement in violent political revolutions (Cuba, Yemen, and Congo) during the decolonization era. Then I turned my attention to North Atlantic Treaty Organization (NATO) connections to the Portuguese armaments used to contain revolutionaries in Angola. In the early 1970s, my focus switched to the prolonged and devastating drought in the West African Sahel. I continued to investigate droughts and looked at the interactions resulting from "people versus nature." I discovered almost immediately that people were using nature to dominate, if not kill, other people. So these kinds of issues do come up, even for those of us who do not belong to the security community. (E.g., United Nations (UN) Secretary General Ban Ki-moon recently suggested that the conflict in western Darfur was the first conflict related to global warming, an erroneous suggestion in my view.)

This brings me to the question of definitions. If you go to the Old Oxford Dictionary, you will find "security" defined in all kinds of ways. First and foremost, it carries the connotation of protection from harm. This can be many things—not just political, but also cultural, economic, or environmental. What we want to do is to find some way to increase environmental security. Picture Humpty Dumpty sitting on a wall, and Humpty Dumpty is about to have a great fall. We need to figure out a way to stabilize that egg by making some kind of constraint. An early warning system is important to this effort to reduce its insecurity.

The Climate and Security Problem.

The climate-related aspect of security is not a new problem. Once people began to populate the Earth, they devised ways to secure enough food and water for survival. They also figured out how to cope with variations in precipitation and temperature from one year to the next. Humans figured out pretty quickly how to adapt to the vagaries of climate. For example, those who insisted on planting, say, corn in the winter in Iowa did not survive; those who waited until springtime to plant were able to survive. So governments are security-oriented, so too are individuals. When it comes to food security, most people have heard about the Genesis Strategy.

There are countless examples of how weather and climate extremes have affected military conflicts over the centuries. There is Napoleon's famous attack and retreat from Moscow in 1812 when only about 22,000 of his half a million men survived because of the unexpected early frigid conditions. Most people do not recall the French victory at Texel in Holland in the harsh winter of 1794 when the French cavalry captured the Dutch navy because its ships were frozen in ice. During the Iran hostage rescue attempt in 1979, dust storms played a major part in undermining the operation because they grounded the rescue helicopters. And in October 1998, El Niño-related rains in northern Peru and Southern Ecuador caused a slowdown in the hot border war between them, allowing time for peace discussion to take place. A local Ecuadorian headline noted: "El Niño Impedes Conflict." Because of the cloudiness, Peruvians pilots could not see where they were to bomb; and Ecuador's army tanks and other vehicles were bogged down in mud. The hot war ended thanks to a timely El Niño in the Tropical Pacific. That climate can affect security is not a new revelation.

Because the global climate system today is changing, we face a variety of different types of threats to climate-sensitive sectors of society and a wide range of human activities, some more menacing than others.

Early warning systems (of which monitoring is an important aspect) are critical to the very survival of states as well as corporations and human settlements. They exist everywhere but do not always work well. This is due to a complex array of factors, among them a failure to accurately forecast extreme weather events and, perhaps still more importantly, a failure to listen to warnings. Another problem is funding for such monitoring aspects. In the absence of hazards, governments and societies lose interest in providing support for them, and they wither.

The Hazards.

Climate is changing, indeed, the global and regional climate systems are changing. There are climate extremes such as droughts, floods, fires, and tropical storms going on around the globe all the time. Today's climate system is not like the one that existed a thousand years ago. Take 2001, for example. This was not a particularly spectacular year. In fact, it was a La Niña year, which usually is seen as a "we do not care" year. But El Niño gets mentioned on the Jay Leno or the David Letterman show or on Entertainment Tonight, while La Niña, El Niño's counterpart, gets no attention. But it is very worthwhile watching. In fact, we ought to care because La Niña events, for the United States, may actually be more damaging and costly per event than El Niño. Besides El Niño events, climate change is likely to make anomalies of climate, water, and weather occur in more places, with greater intensity, and more often.

It is important, though, to keep in mind that climate is not the only thing that is changing. Societies are changing as well. It is vital to remember what physical scientists tend to forget—that people are, unlike a thousand years ago, now an integral part of the climate system. People are doing things that have an impact. What we do at a local level matters as well: We know there is such a thing as the urban heat island effect. What we do regionally matters; e.g., if you deforest parts of the Amazon, it changes rainfall patterns within the Amazon basin. And what we do globally matters, such as burning fossil fuels, altering wetlands, maintaining feedlots, tropical deforestation, and the like.

Creeping Changes. With regard to hazards, there are quick onset hazards and there are slow onset (creeping) hazards. I have been focusing on and writing about the creeping ones. The creeping ones are those that eventually bite you when you do not expect it, so you have to be aware of them. Most environmental changes in which human activities are involved are of the creeping kind. Unfortunately, creeping changes have unknown thresholds for unwanted step-like changes. You can recognize those thresholds once they have been crossed and a crisis has developed. At that point, however, it will probably be more expensive to deal with. A rich country can somehow cope. A poor country, however, has to live with the consequences of the problem. In both rich and poor countries, it is very difficult to maintain monitoring systems for environmental changes that are of the creeping kind, that is, long-term, low-grade, but cumulative.

One of the biggest problems facing all societies in a warmer climate regime is health in general and vector-borne diseases specifically. The climate-health issue is now starting to get a lot of attention because malaria, dengue hemorrhagic fever, and the hanta virus are moving into Texas from Mexico. It will likely not be long before these types of tropical diseases make their way into other border states. This hazard is real, along with other creeping hazards such as deforestation, soil erosion, ozone depletion, global warming, CO^2 emissions, SO^2 emissions, urban sprawl, and coastal erosion. In the early 1980s when I was on climate impact advisory committees for the UN, the committees would list 10 issues for financial support. Food, water, and energy always made the top of their list; then fisheries. And, finally, at the bottom and of least interest was public health and public safety. By the time they finished doling out support for various issues, there were never any funds left for either fisheries or human health.

Abrupt Changes. As noted earlier, there are two types of change, creeping and abrupt. Now I am a tree-hugging liberal and believe that human induced global warming is underway. However, some of the things that scientists say, and the media repeat, do worry me: Talk about abrupt change is one of those things. In the 1970s there was a lot of hype about how the West Antarctic ice sheet was going to fall into the ocean and raise sea level precipitously by about 8 meters. This is what could be called a "dread factor." Scientists continue to looking for a dread factor that would convince society that it must act quickly. This does not mean that abrupt climate change is not likely to happen. It can. But it is not clear that is the best way to garner support for taking action to combat global warming. Personally, I continue to be more worried about creeping changes.

Early Warning Systems – Key to Survival.

There is actually no such thing as "an" early warning system. There are actually lots of early warning systems in place within any given country, and often some of these systems are focused on the same problem. Warning systems are more important than many or possibly even all governments realize. We tend to think of societies as stable when actually they are not. The image that comes to mind is that of a pyramid in which the base represents society and the apex represents a country's early warning systems. It operates like a searchlight shining around the base of the pyramid, ever vigilant. The reality is that the pyramid rests on its apex, which means that a society is in unstable equilibrium and what stability there is rests precariously on the apex, e.g., it rests on its early warning systems. This is as true for rich industrialized countries as it is for poor agrarian ones. Ultimately, whether or not these societies are able to survive depends on whether or not they have adequate warning systems.

A warning system is composed of much more than just the component that prepares a warning based on its monitoring activities. It is a *system* and has many parts. The system encompasses monitoring processes for a hazard, effective dissemination mechanisms for the warning, response mechanisms, and reconstruction efforts. It also includes constant feedback to those responsible for monitoring. A key point to bear in mind is that a warning is useless if no one is there to hear it. Information alone is not enough. It is like clapping with one hand; there is no sound. Again, if an early warning system is to work, someone must receive it, understand it, translate it, disseminate it, and use it effectively.

Warning Failures. Warning is about forecasting and also about people *listening* to forecasts. Warning sometimes fails because science lets us down, and it sometimes fails because people do not respond to information given them. All this can be attributed to a complex set of factors. Among them, we may note the limits of our scientific understanding, the uncertain nature of the information that reaches us, government dislike of monitoring, fear of cost, a widespread failure to understand the fragility of societies, a tendency not to take responsibility (knowing where "the buck stops"), ignoring problems that do not obviously or directly affect something we care about, and a failure to correctly prioritize problems.

Forecasting and Foreseeing. What, then, is the scientific community's ability to forecast extreme or high impact climate, water, or weather events? By example, El Niño events (and their correlated socio-economic and environmental impacts) can provide insights into both the kind of impacts that might result from global warming and, more importantly, how prepared societies are to cope with such impacts. Although there is not a one-to-one correspondence to global warming's potential impacts (e.g., location, intensity, and frequency), El Niño research can serve as an interesting glimpse of a possible future for some regions of the globe. I believe in "forecasting by analogy," having studied El Niño and its impacts and climate change issues since the mid-1970s.

The scientists watch out for the onset and development of an El Niño by monitoring the Pacific Ocean at the surface and below, using technologies like hi-tech buoys and satellites. Anyone can monitor these changes now on a daily basis on the Internet if they wish to do so, given advances in our environmental monitoring capabilities. Graphs in the Internet Google search shows what La Niña and El Niño conditions look like. There are also many variations of impacts maps. However, the original maps on which those impact maps are based have not been updated since the late 1970s! So, they must be used with great care as some El Niño-related impacts have changed over time. There have been seven El Niño events since the late 1970s and, despite the heavy dependence of decisionmakers on them, researchers have been unable to get even a proverbial "dime" to update them. The point is that we use information in our decisions—and not just some existing El Niño information—that masquerades as scientific data. But we must keep in mind that climate is constantly changing naturally even if there is no human influence on it (which there is) and what was useful information 10 years ago may not be all that correct or useful now.

People have forgotten about the hurricane season of 2004. However, it was an important year in terms of climate-related impacts. A record-setting four hurricanes made landfall in Florida. Before that, three had been the maximum number. In the Pacific, ten typhoons made landfall in Japan, before which the record was eight. 2005's hurricane season was notable and record setting: 28 named storms formed in the tropical Atlantic and Gulf of Mexico. It included the devastating Hurricane Katrina as well as notable hurricanes Rita and Wilma. The 2006 hurricane season was surprising for a different reason: Pundits forecast another harsh hurricane season when it turned out to be a very mild and uneventful one. I do not know what was going on, but more importantly and unfortunately, the scientific community also does not know.

Foreseeability. "Foreseeability" is a word I borrowed from the legal profession to apply to climate forecasting. Though I am not a lawyer, I believe the concept has value and could be used by impacts researchers and for those who seek to communicate about climate change and its impacts to the public and policymakers. To me—a social scientist working amidst physical scientists—foreseeability can be viewed as a qualitative version of probability expressed in statistical terms. For example, suppose that I am driving a car in a location I have never visited before. When I come to a stop sign, I can stop at the sign or take a chance and drive past it. I have no idea whether there is a 10 percent probability of getting broadsided by another car driving on the cross street, or 50 percent, or 80 percent. Even though I do not know the probabilities of being hit by another car, it is foreseeable that a car could hit me. I do not need more science (that is, a better probability statement) to understand the situation at this particular stop sign. An accident is foreseeable.

Apply this to climate change. Here is what the scientists are saying about super storms and about high impact climate, weather, and water events: super storms *are* bigger in intensity, magnitude, or frequency. Everything is being labeled "super" for a good reason. Some storms are super because of their physical parameters: winds are stronger than ever, the sea level is higher than ever, the temperature is record setting, sea level pressure is the lowest on record, the rain fell in greater amounts than ever, etc. We may even have several superstorms within a season. Meanwhile, as the climate changes, our ability to forecast it deteriorates. That is obvious when you think about it, because the history of the past is not as useful for projecting the future if the climate is changing in a major way. Hence, foreseeability can be a useful concept.

Clapping With One Hand. As I said earlier, warning is only useful when the warning is heard, and when the warning is understood by society at large and especially by those at high risk. With regard to understanding the needs of society for early warning systems, Hurricane Katrina provides a "teachable moment." Whether we are looking at how governments respond to crises, or how humans respond to warnings, a review of Katrina can provide insights. The forecast from the National Hurricane Center on August 26, 2005, noted the high probability of landfall near or at New Orleans. The situation within the hurricane's "cone of uncertainty" looked bleak. At that time, the director of the hurricane center advised policymakers, "Evacuate people." He felt confident in his forecast. It was 60 hours in advance. That is a respectable lead time to evacuate. However, forecasts, as we now know from the outcome of this event, are simply not enough to save lives or property. After Katrina, a respected scientist noted that "We were able to provide 60 hours of warning with our WARF model. Give us more hundreds of millions, and we'll get that reliable forecast out to 70 hours." The response was that society would likely be as unprepared to cope with a 70-hour lead forecast as with the 60-hour one. What would be needed would be funding for research to better understand society's ability to *use* such forecasts with respectable lead times. The best value for the money would be to develop other sides of the early warning system, not just improve the science component.

Despite the successful 60-hour forecast, four levees collapsed. The images of people, environment, and impacts were incredible. Despite the heroic efforts of some organizations, the Internet and nightly news exposed a confused helter-skelter response to assist victims along the Gulf Coast with a major media focus on New Orleans. As I watched the news and reviewed numerous photos, I could only ask myself, "Is this America?" Sadly, I watched tens of thousands of refugees of American origin on American soil. I have seen the impacts of horrible disasters in the past such as the images on the Internet from Honduras as a result of Hurricane Mitch in 1998. An estimated 17,000 Hondurans died, and countless numbers of environmental refugees were created. I never expected to see in the United States this kind of dislocation, devastation, and political confusion. A Spanish newspaper reported the following: "America is a Third World country. It responded badly." The United States did well with its rapid military deployment to assist tsunami victims in Southeast Asia. It was fast. Why not here in America?

Making a bad hurricane story even worse, in 2004, a year before Hurricane Katrina, the Federal Emergency Management Agency (FEMA) held a workshop with over 250 government people to discuss Hurricane Pam, a scenario in New Orleans, with just about every relevant local, state and federal agency represented. The Hurricane Pam scenario was based on a category three hurricane hitting New Orleans head-on. The agencies identified what the impacts might be on the population, on the built up environment, on debris removal, on needed emergency medical facilities, on coordinated responses among agencies, on communication needs, and so forth. A year later, the physical aspects of the scenario came to pass in a hurricane called Katrina. It is not clear at all what aspects identified from the Pam exercise were used, if any. Probably some elements were used, but basically the Hurricane Pam scenario was of little value. Those present knew that the levees which were in need of urgent repair were not going to be fixed.

Assuming Responsibility.

There are lots of reasons why we do not always respond to problems as we should—some of which have been discussed already. In part, a lack of appropriate responses has much to do with a failure to accept responsibility for one's actions and perhaps a failure to care if we personally are not affected by the adversities associated with an event.

There are three kinds of responses to risk. There are those who are risk averse, those who are risk takers, and there is a third overlooked kind—risk makers. The risk maker is the one who makes decisions that muddy the water for someone else downstream. They are not affected by the adverse consequences of their decisions. They go back to the drawing board to make other decisions, while those affected have to live with the negative impacts caused by the risk maker. That is something that those who monitor the impacts of decisionmakers must watch out for. Unfortunately, there are more risk makers than one might expect. They must be made to take responsibility for their adverse impacts because they are not reluctant to take credit for their positive decisions.

People everywhere are confronted by lots of climate, water and weather related hazards. For example, recently China has had to deal with its dust storms. Mumbai (India) has suffered major intense rainfall and resultant heavy flooding. Australia has had to contend with bushfires threatening its cities. Problems such as these (droughts, floods, fires, vector-borne disease outbreaks) will continue, and it is foreseeable that they will increase in frequency and intensity, and appear in locations where they had not occurred in recent times.

In the early years of the 1990s, it was politically incorrect to talk about climate-change-related winners and losers. When you start talking about winners and losers, you are implying that someone has caused a problem, and, if that is the case, then that someone has the responsibility to address the problem. With regard to climate change, the U.S. Government, among others, has not yet accepted its responsibility.

There is a well-known dilemma that comes to those forced to make decisions under conditions of uncertainty. I would guess that all societies have adages that are meant to guide people's actions. Sometimes those adages send conflicting messages: "He who hesitates is lost," but also that "You should look before you leap." The European Environment Agency (EEA) has addressed the "decisionmaking under uncertainty" dilemma by fostering consideration of the "precautionary principle." Its publication is entitled "Late Lessons from Early Warnings." To understand this concept, think about our politicians' response to asbestos, one of the EEA's case studies. In the 1890s, scientists say that asbestos in your lungs can kill you, and political leaders respond by saying that there is not enough science to support the claim. Again in the 1920s, "asbestos in your lungs will kill you," and again the political response—"We need more studies, not enough science." Yet again in the 1960s, "asbestos in your lungs will kill you." This time there was too much scientific information to support that view. The political response was "Oh, let's do something. Let's rip out all the asbestos we put in from 1890 to the 1960s." Global warming, it seems to me, is one more example where the warnings have been vocalized for more than a century but which have been ignored.

In 2001, I organized a 16-country study of the impacts of and responses to the 1997-98 El Niño. The book that followed was entitled, *Once Burned, Twice Shy; Lessons learned from the 1997-98 El Niño*. While preparing the summary of the project, I came to realize that lessons associated with El Niño-related disasters are really only *identified*. They are the same lessons that had been reported in other countries as a result of all kinds of hazards. Since then, I have come to realize that in most disasters lessons are really only being identified and are not really *learned* unless they are *applied* to improve a society's response to future similar hazards. I have come to believe that the very phrase, "lessons learned," is part of the problem, because when we say it, we think someone is taking care of it when, in fact, no one is.

Conclusion.

In short—we have been warned and should heed the warning. Climate is changing, but climate is not the only thing that is changing. Society is changing as well. As a last thought, I want you to envisage Shanghai as it looks now, a modern city of the 21st century, as opposed to Shanghai as it looked just 2 decades ago, an industrial city on one side, the river and a marshland on the other. When it comes to rapid changes along the coast or in an urban center, Shanghai is by no means alone. Without taking into consideration environmental changes and especially climate change (a.k.a. global warming), governments in most countries will continue to allow their citizens to move into harm's way. The climate-change future that people fear seems to be arriving earlier than expected!

Good Governance and Stability

Robert H. Dorff
Strategic Studies Institute

This panel, as a whole, was asked to take a look at what we can do to prepare for a world changed by global climate change. I myself was asked to address what the international community in general and nongovernment organizations (NGOs) in particular can do in this area. How, for example, might they promote the kinds of institutions that might act as a break on potential instability resulting from global climate change? Some fascinating questions have been raised earlier today: "What causes climate change? How should we address it? Will climate change cause conflict requiring national security responses? Is climate change a security problem?" I do have views on these issues, but I do not plan to take sides in these debates. They are not essential to the argument that I am going to present today. My argument begins with the following:

First of all, it is my view that *the* strategic imperative of the 21st century for the United States, and indeed for the larger global community, is the persistent absence or weakness of effective legitimate governance and the corollary absence of the basic functioning market economies that generally go along with that. For reasons we can discuss later, I deliberately avoid using the term, "democracy."

In so saying, I am also directly implying that several of the persistent challenges we face—terrorism, transnational organized crime, insurgencies, repressive tyrannical regimes (what we used to be able to call "Rogue States," or "axes of evil," but now are officially called "States of Concern")—stem largely from this overall persistent weakness or absence of legitimate governance. And because of that, a U.S. grand strategy (encompassing our national security strategy) should have, as its core objective, the promotion and sustainment of effective legitimate governance and the market economies underlying them.

Another point in this argument is that such an overall strategy must address the potential challenges, threats, and so on that can result from global climate change. Here, regardless of whether or not one is an agnostic on these other questions, the United States clearly has a very important role to play. In other words, addressing global climate change makes good sense as part of helping implement this overall grand strategy and U.S. national security strategy.

I will also argue that strategic success will depend upon civilians working together with the military, to include U.S. Government civilian agencies, foreign government civilian agencies, foreign militaries, NGOs and even—indeed especially—the private sector. This kind of cooperation could go a long way towards addressing some of the problems created by global climate change.

Let me try to connect a few of the dots. First of all, one of the themes that leapt out at me yesterday as I listened to the discussion was this notion that global climate change factors are at least stressors, if not direct causes of conflict. Many presenters pointed to the fact that no matter our perspective on global climate change, its impact will vary. It will vary, in part, because of some geographic and environmental "givens," such as where affected countries and people are located. There are, quite frankly, some things which we really cannot change, short of moving entire continents with some wave of the magic wand!

However, sometimes the impact will vary because states and people have different capacities to respond to climate change. That is to say, they have different abilities to adapt and change before the consequences are felt and different abilities to take measures to mitigate the damage once it is done. My argument is anchored in the assumption that we do have the ability to change these kinds of capabilities. It may not be easy, and when what we do about it comes to promoting effective, legitimate governance and increasing the effective capacity of organizations and especially of states, there are obviously some very, very difficult challenges ahead. But I think that, in most strategic interactions, you always have a couple choosing. One is to do nothing, and another is to go one step further and do nothing with a lot of flair. I call that preemptive surrender. Preemptive surrender—choosing to do nothing in advance of something you are pretty sure is going to happen—is clearly something that we can do in this whole area of global climate change. It is also something that we can do in the face of the persistent weakness or absences of legitimate effective governance. We can choose to ignore the strategic imperative. I do not think this a good strategic choice, and I think the consequences of it are very, very serious.

Let us continue to connect the dots and pick up on some of yesterday's themes. Again, humanitarian and natural disasters may not, and I believe do not, cause states to fail. However, failing or fragile states generally *do* cause the impact of those disasters to be worse. It is, I think, scientifically shown that, though you may not be able to prevent the famine from happening, the impact of famine is much greater in those states and regions where institutions lack the capacity to deal with it.

Dr. Price-Smith commented that disease may or may not cause national security threats or conflicts. System capacity, he said, is a key intervening variable. And that is fundamentally the argument that I am making here. Infant mortality is a *predictor* of state failure. But you are not going to address the problem of state weakness and state failure by reducing infant mortality. You are going to reduce infant mortality because you are going to find ways to address the capacity problem of the state or the region or the community. This has policy implications. And part of our charge here over these 2 days was to think about what we can do at the policy level, rather than simply coming back and raising questions and more concerns.

So let me take a look at security, stability, transition and reconstruction operations (SSTRO). The military clearly has a role to play. This is especially true when we are trying to move countries that are in conflict into post-conflict situations or reconstruct countries in a post-conflict situation.

In almost all of these kinds of situations, what a Venn diagram would show is that you have a humanitarian space and a battle space. The focal points for a lot of these SSTRO are to be found where those two overlap. My view is that this is where we need to do a lot of work. The different organizations, agencies, and people (civilian and military) have different capabilities. We need to try to figure out how to synchronize, harmonize, and coordinate the work where they overlap so that we bring the best capabilities from the right places to bear on these problems.

Combatant commands, as Kent Butts noted yesterday, are already doing this. They are reaching out and co-training with and co-deploying with civilians. This is a good thing. However, it should be driven by overall policy and strategy, rather than something that the combatant commanders have to make up as they go along because they have been given the military policy guidance to go forth and do these things. We have yet to get this right.

I would also like to address one related issue that surfaced. Several persons objected to imposing so many tasks on the U.S. military. How, they asked, can they, the military, continue to do everything they are asked to do? And should they be the "world's policeman?" In response, I would say, that no, they should not be. That said, why can we not tackle global climate change as part of an overall problem of promoting and sustaining effective legitimate governance around the world? And why do we not do this with all of our resources and capabilities, civilian **and** military?

It seems to me that global climate change response capabilities, as well as global climate change consequence management and mitigation, will reside significantly with state and local political capacity. I do not think it is a matter of creating more large, sweeping international organizations to tackle these things. Our task is to break this down into "doable" parts and pieces. It does appear as if the worst disasters are likely to hit in places where much work remains to be done—and not necessarily as much at the state level as at the community level. There is much that is local in building good governance, and much at that level to be done about global climate change.

To the extent that we can integrate global climate change response, consequence management, and mitigation capacity building as components of promoting and sustaining effective legitimate governance, we have a very useful convergence. This is my core argument.

In trying to promote legitimate governance and democratic norms of cooperation and negotiation in post-conflict environments, it is useful to work at very local levels. Take, for example, the building of a school. If you involve the parties to the conflict in the design of the school, you can create the equivalent of a parent-teacher's organization and possibly a curriculum design effort. You get the parties to focus on the fact that this is about their children's future and not about how they do not like each other and never have liked each other. You also get them to appreciate, by the way, that they do not have to learn to like each other to make those kinds of legitimate, governmental institutions work.

Why could we not do the same things when it comes to meeting the climate-change challenge? We could do some of the things that the scientists in this room say need to be done to mitigate consequences as well as do some things to change habits. We could build institutions and norms that will prevent future bad consequences of global climate change while also promoting effective legitimate governance.

There are many groups that could be involved, including nongovernmental organizations and the private sector. I have spoken with Fortune 500 companies about strategy and strategic leadership. If you look at the areas where they are operating, they have a vested interest in creating this kind of effective legitimate governance. And, by the way, many of them are in the business of doing things about mitigating consequences and preventing global climate change. They can also be effective partners and actors in this.

So preparing for a world changed by global climate change is really about bringing to bear the full range of actors with all their capabilities across the elements and instruments of power—diplomatic, informational, military, and economic. I am mindful of the Russian aphorism which says, "If all you have is a hammer, every problem looks like a nail." What we need to do is figure out how to organize and bring to bear on specific problems the most appropriate elements of power. This includes learning how to get the military and civilian organizations to work together most effectively. I think my contacts in the military would tell you they would be happy to see capabilities developed elsewhere and happy to work together with other actors to achieve the goals.

Defense Planning

Henry H. Gaffney, Jr.
Center for Naval Analyses

Introduction.

Across time, I have done a lot of work on conflict in the world. Among other things I have drawn up an inventory of U.S. force responses to situations from 1970 to the present—the most complete one in existence. I have also analyzed the "American way of war" as it has been exercised in the very few combat situations since the end of the Cold War and the global war on terror.

There has been a surprising emergence of an elite consensus on both energy and global climate change. This has, perhaps, arisen only over the last 2 years, although a lot of work dates back to the 1970s. I myself have found the briefings on global climate change by Jim Hansen and Tony Ganetos quite convincing—so long as one can handle ranges of projections off the current data, scenarios of effects and outcomes, and assuming no mitigation and adaptation along the way. So I am not one of the skeptics, that is for sure. I am an innocent bystander.

The first point I would make is that the whole process of global climate change and the associated energy business is going to be evolutionary. I agree with Mickey Glantz on the question of the creeping process. I have heard people talk about abrupt changes, but I have not been convinced. And then I heard yesterday that "abrupt" in some cases is supposed to indicate a period of years. Altogether, it seems more likely that warming is going to creep up on us. Not everything, therefore, will go to pieces at once. And the metaphor which best describes this—this will the first time it is used in this conference, but one that comes up with both energy and climate change—is that of the boiling frog. Is anybody not familiar with that metaphor? No?[1] Good, I will proceed.

Based on what I have read so far, it is hard to postulate much about the wars that are supposed to inevitably arise thanks to global climate change. What gets mentioned are water wars, failing states, that is, failing governments, and mass migrations to cities, across borders, and worst of all, to the north.[2] Among those mass migrations will presumably be waves of terrorists, supposing, that is, that 20, 30, 40 years from now the terrorist problem is exactly the same or worse than the one we have got today. One could argue about that, but that is not what this conference is about.

Ironically, all these conflicts are expected to emerge in places which do not generate much $C0^2$ and which will not contribute much to global warming. I am talking about Africa and the Middle East, especially.[3] Let us remember, though, that any new conflicts to emerge in Africa in the course of global climate change would do so from what is at the moment a primarily peaceful base. Although the U.S. military likes to talk about a dangerous and uncertain world, state-on-state wars have practically disappeared, and internal conflicts have been dropping like a stone. What some of us have discovered in our discussions is that people think that state-on-state wars are more likely to happen when outside states intervene in some internal situation that has gone very bad. Right now we might allude to Iraq in that regard.

Internal conflicts have dropped both by my own count and that of researchers formerly working at the

University of Maryland, but now at George Mason University with Monty Marshall.[4] We Americans are obsessed with the possibility of terrorists spreading like weeds in ungoverned spaces like the Sahel. And some even talk about terrorists building nuclear weapons there despite the lack of water and electricity. That would be an interesting thing for them to try to do. We have got to take some of this with a grain of salt. Last May, I was at Special Operations Command Europe's International Conference in Stuttgart, Germany, where I got pretty thorough briefings on what they are up to in the Sahel, helping the locals to chase what is now called al Qaeda in the Maghreb (AIM). The Sahel is underpopulated for some very good reasons. However, the U.S. establishment can hardly see the whole world now because it is bogged down, physically and mentally, in Iraq, and as a sideshow, in Afghanistan. At the same time, there is a recognition coming out of the wars in Iraq and Afghanistan that conflicts in the future may well involve primarily irregular warfare, responded to by counterinsurgency. I will talk about that a little bit more later.

Worst Case Scenarios.

Let us imagine, nonetheless, that we are living in, say, 2050. Let us further imagine that we have ignored all the evolutions of global climate change, that practically no mitigation and adaptation has been attempted in the interim, and that, therefore, the catastrophes of global warming are upon us. We are faced by drought, flooding, and migrations. Let us further suppose that economies and governance have broken down. The question that I was supposed to address at this conference was "What does the U.S. military do in that eventuality?"

Let us start by revisiting the dilemma of U.S. national security. When we speak of this, we usually mean defense of the homeland. Mickey Glantz said that national security has something to do with protection against harm. We certainly like to defend ourselves far away before threats reach us here. We also talk about stabilizing the world. But in terms of actual shooting interventions, we have done very little of that since the end of the Cold War, and only Korea and Vietnam before that. Now, it's Iraq, Iraq, Iraq.

Through our alliances and deployments of forces, we consider that we are deterring North Korea from attacking South Korea, a Chinese attack on Taiwan, and Iran from attempting to close the Strait of Hormuz. We keep up our strategic nuclear posture. Nobody ever mentions that because it does not have much to do with global climate change. We do not talk about nuclear winter anymore.

The United States spends more than the rest of the whole world on defense and pursues the most technologically advanced equipment. Our belief is that this discourages others from trying to catch up to us. So we have to ask, how do the wars people have postulated as arising out of global warming threaten this national security? Among the threats postulated, the one that comes up the most often is that of mass migrants attempting to reach the United States. If they are from the Middle East, the Arab countries, one concern is that some of them might be terrorists.[5]

The other great threat to national security would be the loss of the imported oil that the United States depends on from the Persian Gulf and from West Africa. Some of the climate change effects—especially the rise of sea levels—could complicate the ability to produce oil in many of these countries. I must say, though, that the problems we have with unrest in the Niger Delta would be solved if the sea level rose and all those oil platforms became offshore platforms!

Otherwise, according to most scenarios, the U.S. role would largely consist of humanitarian interventions, comprising both disaster relief and efforts made to stop people from killing one another. How exactly this can be construed as protection of America is not altogether clear, but why should we rule it out? Humanitarian relief is a good thing to do. We are somehow embarrassed to say it.[6]

By the way, as an aside, none of these evolutions rising from global climate change, with its economic stringencies, if not devastations, are conducive to any country in the world building the big classic technological militaries. They will not have the money and, more importantly, they will not have the need. It is all going to be irregular warfare, as people call it, which means a switch away from reliance on technological capabilities to reliance on massive ground forces.

The U.S. Military Response.

Given these possibilities, how should the United States respond militarily? First, it is not up to the U.S. military. It is up to an administration in office and one presumably working in consultation with Congress, and somehow or other in touch with the public. Many of the articles that I have read about military interventions and collapsing states invoke the passive voice. The authors say things like, "The United States will be called upon. . .," and yet nobody ever specifies who is to do this calling upon. And why do they duck behind the passive voice? Because it can hardly be the UN; after all, we are a member of that organization and would thus be part of the calling. It might be "world opinion" or NGOs or whatever that would do the calling, and we would consider that. By the way, as I was explaining to Kent Butts yesterday, the reason that we got into Somalia was not the CNN effect so much as the fact that the television crews went in there and were able to pay any amount for their protection from the "technicals" in Somalia. At that point, the NGOs were priced out of protection, and they needed the free protection provided by U.S. forces. So any responses are the decisions of the President, presumably in consultation with Congress.

Second, imagine the worst case scenario again. Assume that climate change has not all been evolutionary, happening bit by bit, and that we are suddenly surprised and overwhelmed by it all at once. The prospects are that the U.S. Government and its military would have to occupy much of Africa south of the Sahara and to deploy the National Guard and Reserves all along the Mexican border and the coast of Florida to intercept migrants. And to do any of this around the world, they would have do it without the help of our northern allies, because they will be fighting off migrants trying to come into their own countries. For these purposes, the United States would presumably require enormous numbers of ground forces. As we see, there is a proposal now to increase the ground forces by 92,000 people. I can imagine that if the situations got so bad we would have to reinstitute the draft, but I have been told not to say anything like that because that is too dire a political prospect.

This would call for associated lift to take the forces wherever they needed to go. Just about all would go by sea since no significant tonnage of anything can ever be moved by air. And the forces would have to take with them all the supplies they need, including water, power, and medicines, plus the associated relief supplies since none would be available locally for the needy population. And these forces should then be prepared to stay there possibly for years. After all, we cannot see any way out of Iraq at the moment and this might put us in a rather similar situation.

Once the political decision has been made, it would presumably be largely up to the military to make the effort. As we have discussed in this conference, the U.S. military is the most numerous, best organized, most mobile, and best supplied of all organizations. Moreover, as we know, it is an illusion to think that we would be able to mobilize sufficient people from the other departments of the government. We cannot even do it right now for a single country of only 24 million people.

Preparing for Meet the Threats.

So what does the U.S. military do in the interim to prepare for any of these eventualities—remembering that we are talking about extremes? First, let us remember that the United States itself would not escape the effects of global warming. And this has a lot to do with our coastal installations along the southern coast and the Gulf of Mexico which is where most forces are loaded if they are going to go on an expeditionary adventure.

Second, I have been talking about massive ground forces, which would be ready, trained, and equipped for irregular warfare, massive police actions, and civic action. They would, presumably, in the event of failing states, be out there governing as it were. They would not need to be so high-tech. All of this assumes, however, again, that no mitigation and adaptation had been attempted in the interim and that many bad situations have arisen simultaneously. Much of this, though, I must say, is countercultural for the U.S. military. Right now, they just want to come home from Iraq, bandage their wounds, see their families again, and restore their equipment. And then they are going to be reluctant to get into another Iraq for a long time to come, perhaps for at least 20 years. They might be resistant to going anywhere before that, but again we know that global climate change will not accumulate in such a way as to require it before that time.

The ground forces, at least, have been thinking more seriously about counterinsurgency upon their experience in Iraq. It may lead to a big shift in the cultural orientation of the U.S. military establishment, especially if the possibilities of state-on-state conflict also continue to dwindle at the same time. But the emphasis on counterinsurgency could also fade as time passes, as it did quickly after Vietnam. The whole business of lessons recorded but not learned is a very strong impulse within the system.

Nonetheless, this endless talk now of not being ready for any other contingency because of the quagmire in Iraq really means the forces would like to stay home and simply maintain their readiness. There is also much talk about switching back to major combat operations for the planning and structuring of the forces. The U.S. Congress and the U.S. public would prefer to avoid foreign entanglements. Thus, it is also countercultural for it to think about occupying other countries, especially for prolonged periods. That political problem would arise. But it is true that the contemplation of such contingencies could focus minds beyond the military onto mitigation and adaptation. By focusing on direct contingencies and making it clear how much they dislike having to respond to these dire situations, the military can play a useful role. They can encourage the higher levels of the U.S. Government and the public to take action and take preventive measures. I think that is, in effect, what you heard from Paul Kern last night.

The United States so far has been reluctant to intervene in the messy situation of country collapse. The only place it went in with a major pacification effort of this kind was Somalia in late 1992. This was a political disaster, and the United States has forgone any such other efforts. Iraq was supposed to be a cakewalk. It was not. In the conditions emerging from global climate change, there is no promise of cakewalks. So we will have to undergo a cultural change if we really want to plan for these kinds of interventions in the future. But remember, it is all going to be an evolutionary process. Remember the boiling frog analogy.

Future Scenarios.

In the couple of minutes I have left, I would like to summarize some of the points I have been making by looking at two scenarios: One is on the future of U.S. defense planning and the other is how defense planning might track with changes in the climate picture.

First, when the U.S. involvement in Iraq is over, if ever, the U.S. military could take one of two major tracks into the future. Track 1 is to return to the United States. Administrations would go back to providing for the readiness and recapitalization of the forces, which in turn means a strong impulse in the U.S. military to go back to high technology, which would be to the benefit of the Air Force and the Navy. And what they would like to do is go back to planning for the Great War with China.

Track 2 might involve administrations planning to occupy many other countries or any other country—but as noted, I am not too sure this country would like to do that after its dire Iraq experience. On the other hand, there is a great impulse within the U.S. military right now to stay active in the world through continuous operations out there and continuous engagement with other countries. In that case, the United States would build up its ground forces, stick to fairly low-tech stuff, and pursue the terrorists where it can. And in the interim, administrations would send the military on lesser, shorter interventions of a humanitarian nature and maybe even provide relief in the event of big natural disasters. Track 2 would be more suitable for the climate change scenarios.

Whichever track is chosen, if we suffer another terrorist attack in this country, the administration will want to come back and do homeland defense. There would be tremendous concentration on that function, perhaps to the detriment of U.S. military operations out in the world.

In terms of priorities for the forces, I note some huge gaps between Tracks 1 and 2. If I am trying to occupy other countries and yet return most forces to the United States, there is a huge gap. You cannot do both. Staying active in the world instead of going back to the United States to reinstitute readiness and recapitalization is incompatible. The third gap is pursuing terrorists versus having a big war with China. These take entirely different kinds of forces. The default solution for all this is what I call "mixed priorities"—that is, within a constrained budget trying to do a little bit of everything—maybe not a bad solution, but leaving inadequacies and frustrations across the board within the Department of Defense.

Second, we can imagine a progression from the evolution of the world system through U.S. relations with the world, and then to the role of U.S. forces through illustrative stages of climate change. Let us remember that U.S. foreign policy always lies in between the world and U.S. forces. U.S. forces do nothing unless the administration in office decides to do something (or nothing) with them.

On the first level of the progressive scenario, we are seeing the first signs of climate change, very wispy evidence. It is very difficult to see, but the trends can be projected to some extent as people like Jim Hansen and the IPCC have done. This bare evidence illustrates the difficulty of engaging climate change as an issue at the national level at this point in time.

At the second level, over the next 10 to 20 years the trends in climate change would be confirmed. Let us hope they would not be, but as far as we know they will. Across the same period, we begin to worry even more about reaching peak oil. For the U.S. Government, we will see some slow efforts to do something about restraining CO^2 emissions, a little more work on alternative fuels, a little worry about competition with other countries to get the oil out there on the market on which we would all still be dependent. And for the U.S. military, they will be finally out of Iraq. They will be recovering and there will be debates over counterinsurgency versus major combat operations as the military considers its strategic orientation. But they will be under budget restraints given the twin U.S. deficits—federal budget and current (international) accounts.

The third level down would be when it all goes bad: droughts, governance and economic collapses, migrations, the North raising barriers to migrants, etc. The United States would either be isolationist, that is, retreating to fortress America, or it is going to face the possibility of occupying many countries. That would lead to a huge public debate between those choices. For the military, they are going to have to guard U.S. borders, think about long occupations and nation-building, and consider the most dire domestic political threat of all—reinstituting the draft in order to man these occupations.

But at the fourth level, let us say that we get into an evolutionary situation where we manage lots of cooperation around the world to control CO^2, to work on disasters together, and to help countries to prepare and to stabilize before the dire effects hit them. For the U.S. Government, by that time, we would hope it would be fully into conservation, efficiencies, CO^2 sequestration, and alternative fuels. At the same time, we probably would expect to continue our presence in the Persian Gulf and possibly undertake some limited interventions that could be managed without much loss of life. Maybe the burden would not be so hard as under the extreme circumstances.

To conclude, there is no relentless scenario that is likely to be played out. Climate change is going to be only progressively recognized, efforts by the U.S. Government and other governments to cope are likely to be hesitant and fragmentary, and the U.S. military will stand patiently aside, waiting to be called by the government.

ENDNOTES - Gaffney

1. The analogy—used by Al Gore in *An Inconvenient Truth*—refers to the belief that a frog can be boiled alive if the water is heated slowly enough. Supposedly, if a frog is placed in boiling water, it will jump out, but if it is placed in cold water that is slowly heated, it will not. The story suggests to people that they should make themselves aware of gradual change lest they suffer a catastrophic loss.

2. I fully concur with Robin Dorff about the problem of governance in vulnerable countries. That has been identified as one of the most serious internal problems we have seen since decolonization—especially as old leaders hang on too long, lose touch with their country, and are engulfed by corruption. It always reminds me of the statement by former Prime Minister Victor Chernomyrdin of Russia. In the mid 1990s he said, "We hoped it would be better, but it turned out as usual."

3. My doctoral dissertation was done on Sierra Leone, so I am something of an Africanist with a little feel for what the continent is like. Of course, when I came back from Sierra Leone and went back to work at the Department of Defense, I spent most of the next 12 years on policy for NATO nuclear weapons. It shows you how little interest there was in Africa in the Defense Department back in those days.

4. Monty G. Marshall and Jack Goldstone, "Global Report on Conflict, Governance, and State Fragility," *Foreign Policy Bulletin*, Winter 2007.

5. How all these people may cross the Atlantic or Pacific is not clear, of course.

6. I noticed a statement in a document I read the other day where it said, "National interests do not have to comply with norms." So norms and national interest are incompatible? I have not yet completely parsed that, but it is an interesting thought.

Commentator

William Schlesinger
Cary Institute of Ecosystem Studies

The conference agenda reminds me a bit of the agenda of the scientific group known as Medea during the Clinton administration. Medea was tasked to assess the potential use of the national security assets of the country by environmental scientists. As Medea evolved, it began to focus more fully on what kinds of national assets, presumably reconnaissance assets, could be brought to bear on some impending environmental problems, including climate change.

There is a lot of convergence between what Medea considered and what has been considered here today. When I look at Glantz and Dorff's presentations, I see a number of parallel conclusions. Glantz stressed that we cannot just have science; we need to actually translate science into some kind of policy or action. That is absolutely critical. The best early warning systems are useless unless you do something with them. The subtext of Dorff's address is that you need to have a government that is responsive to what science is saying. If the government represses science, or is ignorant of it, or chooses not to read it, disbelieve it, and instead base their actions on divine inspiration or whatever, that whole scientific effort is wasted.

Hank Gaffney drew attention to a number of potential problems. For my part, I am a little worried that we have not fully considered the cost to the United States of playing a role as a policeman in any future world. I am a bit of a pessimist this way. I tend to think that the United States is going to become a bit like Great Britain and will end up a power that was once grand but no longer has the money to accomplish much on the international scene. I think we will see a rise in the wealth and influence of Asia. And we will not be able to affect their behavior. I think we need to consider what we might be able and willing to pay for.

Let me say a few things in regard to the military and the need for early warning systems. There will always be flat earth people who say that climate change is a hoax, but I think we can say that climate change is an established scientific fact. If we ask what kinds of specific things the military might be called upon to respond to, I would put famine at the top of the list. I think there is every indication that even if the rainfall patterns do not change, if we have warmer temperatures, there will be drier surface conditions and less soil moisture for crops. This will be especially true in the centers of continents, in places such as our Great Plains and Midwestern states that grow so much of the world's cereal crops. The military should be ready for not just local famine but wide-scale increases in the price of food and crop failures. In the face of crop failures, will we be able to feed as effectively as we have been the 6 to 10 billion people that will be with us then?

I would also put disease high on the list. I think we can expect that, in a warmer future world, diseases that are now confined to places that we visit on safari may very well be with us here at home. And, increasingly, we may see people coming home from faraway places bringing these diseases with them. The medical community needs to be prepared to deal with this but so, too, should the military. An early warning system would indeed be useful. In the late 1990s, scientists proved able to forecast the outbreak of Rift Valley fever in East Africa some 60 days ahead of time by a satellite monitoring of sea surface temperature in the Indian Ocean. It would surely be useful to know when one might have to beef up the public health infrastructure overseas or even send the military to stabilize much of East Africa.

There is probably very little that I can say about water in general that has not been said in the popular press or said here over the last couple of days. I will just reiterate that when one looks at which resources are going to be in short supply and that we might think about fighting over, the prime candidate would probably be water.

I spent all my time dodging the draft and avoiding getting in the military back in the Vietnam era when I had hair and a ponytail and all that good stuff. But if I had embarked on a military career and if I were in the Navy, I think I would certainly be concerned about sea level rise. This clearly will have an impact on coastal facilities and ports, and affect the ability to land in various places. In the future, if we are still using submarine warfare, an ice-free Arctic would put to rest the whole cat and mouse game of hiding under the Arctic icepack.

If I were in charge of Army-type forces, I would be worried about the great vast areas of Siberia and Canada that are frozen much of the time now. We have devoted a lot of energy to studying how to fight on frozen ground. And while it may be cold, it is a whole lot more convenient to be on frozen land than in the swamp of melted permafrost. So again, I would say we are poorly prepared to think about any scenario that puts us in a ground war at high latitudes.

As to the Air Force, it would not get off the ground if we did not have low density liquid-based petroleum products. You cannot fly big transport planes with coal. And I think the Air Force needs to give some serious thought as to how we are going to move people from place to place in this future world. Some elite jet fighters could probably operate using traditional fuels. But how would we move massive forces if we have only a limited supply of liquid fuels? At any rate, how would we do so at a cost we might be willing to pay?

I have one further comment. I think we need to fear proposals for global engineering, spraying sulfate aerosols into the stratosphere, and putting mirrors on satellites surrounding the earth to reflect the incoming radiation of the sun. We want to be very, very wary about manipulating the climate with global engineering. As an environmental scientist, that just makes me shudder. Let me stop here and, hopefully, we will have some good discussion.

Discussion

Q: Can you comment on the utility of the kind of indicator-based methodologies and indicator lists developed throughout the Cold War by intelligence organizations?

Glantz: A lot of groups do develop indicator lists to help warn of things like famine. Lists are fine, but they are dangerous. They lock you in. If you stick to what is on the list and do not keep an open mind, you can miss things. Some lists are also overly long. I have seen famine early warning indicators with 40 indicators, including measuring arms and heads and body weights and stuff. Do you need 40 to know that you are heading towards a famine? One big problem is that famine is a creeping thing that occurs over a relatively short period of time. It is also an event—when we learn there is a famine, it means people are already dead. Dead people are not good indicators, okay? You need to look at things like, are people migrating? Are people selling their jewelry in the marketplace? You are likely to get your earliest warning from some truck driver coming into Nairobi telling you he saw folks selling their pots and pans. And you need to heed this and check it out; it is not going to be the sort of thing on an indicator list, however.

Dorff: Clearly we need early warning systems. But failures are often not the result of a warning failure. They happen because there is no political will to act. We need to develop this will, and translate science into policy, and fund the policy, or do some of the things I talked about. As to indicators, yes, these are good. Apart from anything else, they could help us mobilize NGOs and the private sectors and encourage them to see to the kind of tasks for which they are best cut out and do the things that will make a difference.

Gaffney: I want to return to Glantz's earlier point about warning and response. We keep thinking of warning as a bell going off. But it is a dynamic relationship between things that happen and things that you need to do as a result. When you lose track of the Japanese fleet, you behave differently at Pearl Harbor than you do when you know where the Japanese fleet is. If there is a hurricane in the middle of the Gulf of Mexico and it is heading north, you do different things in New Orleans than you do when there is no hurricane in the middle of the Gulf of Mexico. When the buildup of greenhouse gases portends a future of significant change over a great time, you behave differently. You will never get "the bell," except in history.

Q: What is a hydropole?

Glantz: That is the best question of the conference! This is a new way of responding to flooding being developed by the Dutch. The Dutch were always looking at the North Sea, and all at once they got flooded from behind. They have decided to pump billions of dollars into building on rising waters. They are giving space to the river and giving up on levees. As a technology, this may not be readily transferable—it is easier to move 10,000,000 Dutch people and find some place for them to live than it is to find a place for, say, 160,000,000 Bangladeshis. The point is that the Dutch are selling us levee technology to use in New Orleans that they are giving up on. We are not asking ourselves the right questions. Can we protect New Orleans against super storms that come twice? Remember, Katrina was a super storm only in one respect—it hit a city that is half below sea level. It was a category 3. It was not a super storm with big winds like Mimi or super cyclone Arissa. Here is an example of a city that is giving up. Let me also make an observation about mega cities on the coast. They will also face similar problems—storm surges, rises in sea level. Note that because the sea level goes *up* a foot does not mean it goes *in* a foot. When the winds are there and the storm surge is there, it can go in miles. So I think that is really important.

Glantz: And the second part of your question (which you were going to ask!) was on disaster diplomacy—can a disaster be used to bring enemy states together? It has done so on at least one occasion, easing relations between the Turks and the Greeks. One had an earthquake and the other offered to help and did, and then it worked the other way months later. When it comes to Cuban-American relations, the findings are interesting. Cubans and Americans work closely together when a specific hurricane is coming at both of them. Cubans are glad to get training in how to forecast. They come to Miami, and sometimes we even go to Cuba. But if you want to talk about global warming, you cannot talk to the Cubans. Facing a threat is not enough to bring about a long term relationship.

Q: Can we adapt existing security relationships (e.g., with Japan, NATO, South Korea) to these new contingencies, provide value added, as it were? Can we use this kind of new world of crisis as a way to build relations with potentially adversarial relationships?

Dorff: The simple answer to both questions is, "yes." If we can get potential future adversaries, let's say China, to engage with us on some of these problems, we might become positive rather than negative (in the sense of military) competitors. Building some relationships through these means might have all kinds of other positive outcomes. That does not necessarily guarantee we will hold hands and sing "We Are the World" together. But focusing the scientific community across borders would mean a lot. In terms of theater cooperation, it is interesting to note the military-to-military engagement Africa has on environmental issues. The cooperation was not driven by an environmental agenda but by a desire to show militaries in other countries how to have democratic military institutions that were not all about fighting each other or about repressing their own people. As to whether or not this adds value to existing relationships, absolutely it does. I think to some extent, NATO has evolved in that direction already. It is doing some of this. With more encouragement from the United States (call it modeling behavior), there are potentially a lot of other organizations, political and military, as well as other countries that we could engage in this sort of thing.

Finally, I think that if the United States clearly articulated a grand strategy in which it was made clear that our objective was to promote effective legitimate governments, we would have a better story to tell around the world. This would not preclude our retaining a counterterrorism component and other elements of good traditional security. It would just make it clear that we want to help create long-term solutions rather than just play Wack-a-Mole and send our military every time a problem surfaces.

CHAPTER 6

OPPORTUNITIES OR OBSTACLES FOR THE MILITARY

Chapter 6 focuses on how concerns over climate change impact the U.S. military. What can the Army do to promote responsible stewardship of the environment while ensuring combat readiness? What is it doing? Mr. E. Thomas Morehouse suggests that some remedies for climate change may at the same time enhance our security and help meet our energy needs. Mr. David Sheets shows us what the military is currently doing to foster renewable energy, and Dr. Karen Hulme analyzes environmental law and its impact on combat operations.

Climate, Energy, and Security – A Related Set of Challenges

E. Thomas Morehouse, Jr.
Institute for Defense Analyses

Introduction.

Good afternoon. Over the past year, I have been involved in two studies related to the issues of climate change, energy, and national security. The report of the first study, "National Security and the Threat of Climate Change," was prepared by the Center for Naval Analyses (CNA).[1] Sherri Goodman, who served as the Deputy Under Secretary of Defense for Environmental Security during the Clinton administration, brought together 11 retired 4-star and 3-star generals and admirals as a Military Advisory Board (MAB) to examine the current state of climate science and provide their best assessments of the security risks it represents to the nation and the world. The MAB considered the threat, likelihood, potential consequences, and timing of projected changes in climate as they would any other issue that represented a potential security risk. The first key finding of the CNA report was that climate change is a serious security risk to the nation. Another key finding was that "climate change, national security, and energy dependence are a related set of global challenges."[2]

A second study was prepared by a Defense Science Board (DSB) task force that examined the energy security of the Department of Defense (DoD), recommended potential remedies, and assessed their value to the nation's energy challenges.

Climate, Security, and Energy: A Related Set of Global Challenges.

The best forecasts that climate scientists can provide have uncertainties in their timing and magnitude, and probably always will. But the trend is unequivocal.[3] The climate is warming, and there are consequences to the ecosystems on which every living organism on the planet depend. Consumption of fossil-derived energy is also increasing globally, and greenhouse gas emissions caused by fossil fuels are also increasing. This is true in the United States and around the world, but is occurring most rapidly in developing countries. China and India are responsible for the largest increases in absolute terms because of their large populations and rapidly expanding economies. Consumption is increasing for both electricity, produced largely from coal, and for petroleum. The global endowment of petroleum is concentrated in a relatively small number of countries. This means that most of the oil we use comes from foreign sources, and the percentage is increasing. China, too, imports its oil and faces the same social and economic risks from supply disruption as other oil importing countries. Conversely, most of our electricity is produced from domestic sources, about 50 percent from coal. China also has a large national endowment of coal. Nationally and globally, energy consumption is predicted to increase in the future as populations grow and more people aspire to a Western lifestyle enabled by high energy consumption. This means carbon emissions are increasing domestically and globally, and this trend is predicted to continue into the foreseeable future.

I will not get into the controversies surrounding climate forecasts that were discussed earlier in this conference.[4] Suffice it to say that if the climate change that we are observing today continues into the future, there will be consequences for our own national security and for international political stability. Most disturbingly, the warming trends are predicted not only to continue, but to accelerate. As General Gordon Sullivan, former Chief of Staff of the Army, said, "People . . . want to be convinced perfectly. They want to know the climate science projections with 100 percent certainty. . . . We never have 100 percent certainty. . . . If you wait until you have 100 percent certainty, something bad is going to happen on the battlefield."[5] Most military planning is done to prepare for relatively low probability, high consequence events. Climate change is different in two ways. It is more likely to result in a high consequence event and, unlike most threats that are singular in their timing and nature, climate change will endure over a long period of time and is global in scope. In addition to its enduring and systemic nature, climate change is also more certain to occur than the threats military leaders have had to plan for in the past.

By its nature, climate change will create new types of security risks. It will both create new security issues and exacerbate existing ones, making already challenging situations worse. The political stability of energy exporting countries will be affected by climate change, and their energy infrastructure, often located on the coast or off-shore, will be particularly vulnerable to more extreme weather. Many oil-exporting nations are net food importers with rapidly growing populations. As climate changes and as agricultural productivity moves in response, the patterns of supply and demand will shift. So, the relationships are interdependent, with effects in one place affecting others. Security is affected by climate, energy is affected by climate, security is affected by energy, and climate is affected by energy.

Security and Climate. Earlier speakers noted that climate change affects the ability of populations to feed themselves, affects access to fresh water, causes the habitats for disease vectors to move, and threatens coastal infrastructures from sea level rise and storm surge. From an international security perspective, marginal states struggle to maintain governance today. This is particularly true in Africa. Climate stresses will make it even more difficult for marginal governments to meet the needs of their populations. This could easily create an increased need for humanitarian and peacekeeping operations, changing the types of missions the military must be prepared to conduct in the future, and the character of the operating environment in which we conduct them. From an infrastructure perspective, the United States is not immune from the effects of climate change. For example, much of the key military infrastructure is on the coast. Imagine what King's Bay would look like after a meter of sea level rise.

Energy and Climate. The best current example of this interdependency that comes to mind is Europe's experience with its nuclear power plants last summer. France, in particular, was hard hit because nearly 80 percent of its electricity comes from nuclear reactors. Many are located on rivers, relying on river water to provide cooling for the reactor. The extreme heat caused the rivers to warm more than normal, in some cases exceeding the maximum temperature allowed for cooling the reactors. Some plants were required to reduce their output. Higher temperatures for longer periods would have forced plants to shut down.

In addition to generating plants, transmission grids are also vulnerable to extreme weather. This is particularly true in the United States and Canada. A case in point was the August 2003 blackout when over 50 million electricity customers in the Northeast, the Midwest, and in Canada were without power because a tree blew into a power line in Ohio. Hurricane Katrina knocked out a number of oil rigs and took refineries off line in the Gulf Coast. In December 2003, the U.S. Arctic Research Commission issued a report "Climate Change, Permafrost, and Impacts on Civil Infrastructure" which cites the risks to significant energy infrastructure in the United States, Canada, and Russia from melting permafrost to include electrical transmission and oil and gas pipelines.

One of the areas of the world predicted to be most affected by climate change and least able to adapt is Africa. According to some estimates, the United States will be importing as much as 40 percent of our petroleum from Africa by 2015. A stabile supply requires political stability in the exporting nations. Climate change threatens to make that more challenging. The establishment of the U.S. Africa Command (AFRICOM) recognizes the need to help strengthen the capacity of governments in the region to meet the needs of their populations.

Security, Climate, and Energy. To accept that climate is affected by energy is to accept causation—anthropogenic carbon emissions are contributing to climate change. This is becoming less controversial here in the United States than in the past. But it is not controversial at all in most of the rest of the world. Even George Bush acknowledged it in his 2007 State of the Union address, "America is on the verge of technological breakthroughs that will enable us to live our lives less dependent on oil." He went on to say that, "And these technologies will help us be better stewards of the environment, and they will help us to confront the serious challenge of global climate change." Some view this as an important acknowledgement by the administration of the body of science linking anthropogenic carbon emissions to global climate change.

Energy is like a master key affecting security and climate. Energy is the foundation on which modern societies and economies function. It is what differentiates a modern lifestyle from an 18th century lifestyle. Access to abundant and affordable energy for transportation and electrification runs the global economy. Major swings in price and availability can send economies into recession, and cause social unrest and conflict. Some view disruption of energy markets as an act of war. A major objective of the foreign policies of developed and developing nations alike has been to assure a free and open global energy market. In the United States, it is a dominant factor in our national security strategy and national military strategy. One of the major missions of the U.S. Central Command

(CENTCOM) is to assure the free flow of oil from Middle Eastern countries. The formation of AFRICOM is clearly linked to predictions that by 2015 the United States could be importing up to 40 percent of its oil from Africa, and to increasing investments in China to develop African energy infrastructure.

There are a number of other security risks associated with our high oil consumption. While they are not necessarily exacerbated by climate change, addressing climate change also addresses these risks. The first is related to the character of oil exporting nations. A number of these are corrupt petro-states that are neither free nor democratic. Many use their oil revenues to subsidize the lifestyles of their populations and maintain their autocratic hold on power. Our relationships with these regimes conflict with our commitment to freedom and democracy, undermining our moral credibility on the world stage. Second, the money these petro-states get from our oil purchases also more directly undermines our interests: some use this revenue to fund terrorists to fight against us. As Jim Woolsey, former Central Intelligence Agency (CIA) director, often says "This is the first time since the Civil War we've funded both sides of a conflict." Finally, there is the wild card of peak oil.[6] Estimates of when we begin to run out of oil range from "we're already at the peak," to "we won't get there for another 40 years." Either way, the debate over the reality of peak oil appears to be over. Given the long time lag for new technology adoption and infrastructure conversion, there is an urgent need for action regardless of which forecast you believe.

Threading the Needle –
Finding the Win-Win-Win Solutions.

The key challenges for this generation are to achieve energy security, national security, and climate security. This is a tall order since they are so interrelated. We cannot act on one without affecting the others, so we cannot think of any of them in isolation. Figure 1 illustrates these relationships.

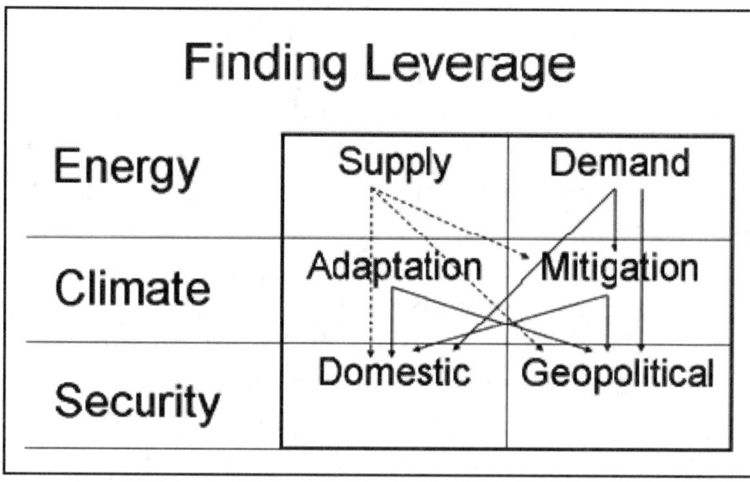

Figure 1. Finding Leverage.

Our security is affected by the decisions we make about how we use energy, the kind of energy we use, and where we get it. Paradoxically, the more fossil energy we use, the more we contribute to the kinds of climate-induced geo-political instabilities that threaten traditional energy sources and the extreme weather that threatens our energy infrastructure. There are two approaches to our energy problem—figure out how to use less, the demand side; or figure out how to get more, the supply side. Supply side approaches to meeting our energy demands can mitigate climate change, but they must be renewable sources that reduce the amount of carbon emitted per unit of energy produced. Supply side approaches that increase the amount of carbon emitted per unit of energy may respond to security or economic concerns in the short term, but increase our longer term climate risks. For example, if we choose to meet our growing electricity demand by building more pulverized coal plants without sequestering the carbon emissions, we accelerate climate change and increase our future security risks. Choosing synthetic fuels that produce higher carbon emissions than petroleum to replace foreign oil, such as coal-to-liquid, tar sands, or oil shale, have the same effect. Conversely, choosing energy paths that reduce our carbon emissions, such as energy efficiency to displace demand or renewable sources, will have the opposite effect. Reducing energy consumption has a positive climate effect.

So, we have choices, and those choices have consequences. Supply side remedies need to be sensitive to their carbon consequences, and demand side remedies should be valued for their contribution to reducing climate risks. For those still questioning the science of causation, there is a pragmatic aspect to consider. There is growing acceptance by both government and industry leaders that the future constraints on carbon emissions are inevitable. The question is not if carbon emissions will be controlled; but rather when and how.

Options for Addressing Climate Change.

Just as there are two approaches to addressing energy: demand, and supply; there are two approaches for addressing climate change: adaptation, and mitigation. And just as for energy, the two approaches are not mutually exclusive—they both have a role to play.

The guiding principle we used in our CNA study was "mitigate when you can, adapt when you cannot." *Dictionary.com* says that to "adapt" is to adjust oneself to different conditions. To "mitigate" is to make things less severe. To advocate mitigation requires an acknowledgement that there is a connection between climate change and green house gas emissions. Mitigating climate changes requires making decisions today to avoid future adverse events. It will require sustained and determined political leadership.

A good starting point for discussing energy options is scale. The United States has 2 percent of the oil reserves in the world, about 8 percent of the production, and about 5 percent of the world's population. Yet, we consume about 25 percent of the world's oil, about 60 percent of which we import.[7]

To reduce dependence on foreign oil is to reduce our foreign trade imbalance, reduce constraints on our foreign policy options, and reduce the funds we send to hostile regimes. As the price of oil increases, there is a corresponding transfer of global wealth from oil consuming countries to oil producing countries. Some advocates propose that if oil consuming countries such as the United States would only drill more, they would import less and be more energy secure. Unfortunately, most independent assessments reject the notion we can somehow drill our way to oil independence. On the other side of the argument, organizations such as the Rocky Mountain Institute (RMI) argue that reducing oil demand by providing incentives to increase efficiency are a significantly less expensive and less risky approach to reducing our oil dependence than either buying more fuel or more drilling.

Others have also come to this conclusion. This suggests that we can make significant progress by focusing on the demand side, or improving efficiency. Efficiency can also mitigate energy price increases on a personal level. For example, if gas is $3 a gallon and you have a car that gets 20 miles per gallon, your gas bill is the same as if you pay $6 a gallon for gas but your car gets 40 miles per gallon. Buying less fuel also provides the kinds of national benefits we discussed earlier: less oil imports, improved balance of trade and global wealth transfer, reduced funding to supports terrorist activity, reduced constraints on our foreign policy options, reduced carbon emissions, and improved long-terms security posture of the country. In short, efficiency is patriotic.

Efficiency also enables supply side options that are beneficial for climate protection. For example, because ethanol has only about two-thirds the energy density of gasoline, a car that gets 40 miles per gallon is easier to power with ethanol than one that requires twice as much fuel.

The same goes for electricity. If through efficiency improvements you can do the same job with half the electricity, it becomes easier to satisfy demand with renewable energy supply sources. This is true for two reasons. First, because renewable energy is generally more expensive than power from a pulverized coal plant, and second, because the technical potential of renewable energy facilities to produce large amounts of electricity is less than coal plants. So, efficiency is an important mechanism to enable market entry for renewable energy. Efficiency also enhances the security of our national electricity supplies. The national grid transfers large amounts of power generated at large central stations over long distances to end users. Renewable energy is produced in smaller quantities and is more distributed. Smaller, more distributed generation is inherently more secure and more efficient than a large system distributing centrally generated electricity.

So, demand reduction measures for both electricity and fuel mitigate climate change and improve security. Supply side measures only do so if they are carbon neutral as a minimum, or preferably, produce less carbon than current sources.

The Role of the Department of Defense.

DoD is a small market presence, using less than 1 percent of the nation's energy. Further, DoD's energy use patterns are significantly different than the rest of the nation. About 75 percent of DoD's energy use is jet fuel used for mobility to power combat and combat support systems. DoD's electricity use is only about 12 percent of its total energy use. For the nation as a whole, petroleum for transportation and electricity are each about half of the total energy demand. DoD does not use gasoline for operational systems, and uses relatively little for leased fleet vehicles at fixed installations. While ethanol can be an important replacement for gasoline as a national transportation fuel, it is of little use to DoD because it cannot replace jet fuel.

To provide some perspective on scale, DoD uses slightly more fuel than a large international airline. In terms of access to commercial fuel supplies, DoD enjoys eminent domain over commercial contracts when needed for national security. Because DoD needs such little fuel relative to the global market and enjoys eminent domain over energy supplies, it is difficult to conceive of a scenario where DoD would not be able to access the petroleum products it needs from commercial sources to perform its missions. DoD's primary petroleum problem is moving the fuel from the point of commercial purchase to the point of use by deployed combat or combat support systems. DoD can contribute to the solution of the national energy problem through its market power. But this power will not come from an ability to create a domestic market for a new fuel source. Rather it will come from its ability to develop the technologies it needs to reduce its operational fuel demand, and the market's ability to leverage those to commercial purpose. Solving the problem of abundant and inexpensive global oil supplies would not solve DoD's most important fuel problem.

Where DoD's Energy Needs and National Energy Needs Align. In August 2006, Marine Corps Major General Richard Zilmer, Commander of Al-Anbar in Iraq, submitted an urgent request for renewable energy systems due to the vulnerability of American supply lines to insurgent attack by ambush and roadside bombs. The request said, "Reducing the military's dependence on fuel for power generation could reduce the number of road-bound convoys. Without this solution, personnel loss rates are likely to continue at their current rate. Continued casualty accumulation exhibits potential to jeopardize mission success. . . ."[8] While Major General Zilmer asked for renewable energy sources, improving the efficiency of systems deployed to forward operating bases turned out to be a more effective remedy to the problem he described. Indeed, the Army's Rapid Equipping Force made far more progress in reducing battlespace fuel demand by implementing efficiency measure than by employing alternative energy sources. There is an important lesson here for the nation.

In 2001 no one envisioned the significant level of force protection that would be needed in Iraq to protect vulnerable supply lines, but this situation has made the recommendations even more important. The DSB report recommended that to achieve this more efficient force, DoD should not allow the logistics burden to be an unplanned consequence of the decision process. Instead it should integrate logistics consequences into its key decisions about force planning and acquisition. As General Omar Bradley said, "Amateurs talk strategy; professionals talk logistics."

To illustrate its point, the 2001 DSB task force asked a rhetorical question: What are the top 10 battlefield fuel guzzlers?[9] Interestingly, of the top 10 deployed systems, only two are combat fired munitions; a tank at #5 and an attack helicopter at #10. The rest were support systems. Collectively, the water heaters for the field kitchens consumed more fuel than the attack helicopters. (See Figure 2.) The task force coined the concept "fully burdened cost of fuel" (FBCF) to assign a monetary value to the fuel demanded by deployed systems. The purpose of assigning a FBCF was to provide a number that the acquisition and requirements processes could use to decide how much it was worth spending in fuel efficiency technologies to reduce the amount of fuel demanded by deployed systems. While this did not capture the operational benefits of reducing battlespace fuel logistics, it did begin to quantify how much force structure "tail" was needed to support deployed forces. Estimates of the fully burdened cost of fuel vary by scenario, and DoD has yet to agree on a standard methodology for calculating it. General Kern, a previous speaker, estimated the cost somewhere between $1 and $400 per gallon, depending on how it is delivered.[10] This is consistent with the findings of the 2001 DSB task force, which worked with the Services to determine the costs for fuel delivery assets. They concluded it cost about $13 to deliver a gallon of fuel to the forward edge of a battle area, and the price escalated rapidly as you moved beyond the forward edge. Fuel delivered by helicopter can cost hundreds of dollars per gallon. In Iraq, fuel convoys are protected by helicopters, driving the cost per gallon quite high because the fuel conveys need so much protection. DoD has made some estimates of the cost of fuel delivery in Iraq, but will not release them.[11] The task force also investigated in-flight tanker refueling—it costs about $26 a gallon to deliver fuel from an airborne tanker if you assume the tankers are free. If you include the cost of recapitalizing the tankers, the cost is about $42 a gallon.

SWA Scenario Using Current Equipment Usage Profile Data

Of the top 10 Army battlefield fuel users, only #5 and #10 are combat platforms.

1. Truck Tractor: Line Haul C/S 50000 GVWR 6x4 M915
2. Helicopter Utility: UH-60L
3. Truck Tractor: MTV W/E
4. Truck Tractor: Heavy Equipment Transporter (HET)
5. *Tank Combat Full Tracked: 120MM Gun M1A2*
6. Helicopter Cargo Transport: CH-47D
7. Decontaminating Apparatus: PWR DRVN LT WT
8. Truck Utility: Cargo/Troop Carrier 1-1/4 ton 4X4 W/E (HMMWV)
9. Water Heater: Mounted Ration
10. *Helicopter Attack AH-64D*

Italics indicates combat systems.
Source: CASCOM study for DSB using FASTALS for SWA.
The end-state force list for SWA (based on the FASTALS Deployment Report) was used as the force structure.

Figure 2. Today's Top 10 Battlefield Fuel Users.

Understanding DoD's Decision Processes. There are three macro decision processes that operate within DoD. They shape the types of missions the Department prepares for, the types of systems it develops and acquires, and the doctrine and operational concepts that underlie its operations. They also establish funding priorities.

The first process, the Joint Capabilities Integration and Development System (JCIDS), anticipates adversaries and establishes requirements for the types of capabilities DoD will need in the future in order to be successful militarily. Its key output is a document called the Initial Capabilities Document (ICD). This document describes the kinds of systems DoD needs to field and establishes their performance parameters. If you think of DoD as a self-contained market, JCIDS is the customer. This customer makes its buying decisions according to a 4-step process: 1) identifying its existing capabilities; 2) identifying the capabilities it needs in the future; 3) performing a gap analysis to decide what capabilities it needs to acquire in order to possess the full capability set it needs in future; and 4) defining how best to fill those capability gaps.

Next is the Defense Acquisition System. It takes the ICD as an input and produces hardware that performs according to its requirements. In reality, there is an iterative process between the JCIDS and acquisition process to match available technologies to the capability needs. Issues such as technology maturity, risk, performance, and affordability are adjudicated to develop an ICD that can be met within the timeframe by which the need for the capability is envisioned. In our market analogy, the acquisition process is the purveyor of products that meet the customer's needs. To meet these needs, the process must develop new technologies, design, and develop new products and oversee their manufacture.

Finally, there is the Planning, Programming, Budgeting and Execution System (PPBES). Its function is to adjudicate the multiyear spending plan and yearly budgets, balancing future year obligations and opportunities with current year demands. Its market function is to provide a source of capital, and assure due diligence over Department expenditures.

Valuing Efficiency — A Military and Security Imperative.

Unfortunately, beyond a must-pay commodity bill in the current budget year, energy gets little visibility throughout DoD's decision processes. The value of more energy efficient operations in terms of military operations, casualties, combat vulnerability, risk to mission, force structure balance, or cost of operations is not calculated, so it cannot be included as a factor in key Department decisions. Even if the information did exist, in the absence of sustained leadership priority, policies, procedures oversight, and accountability, it would not become a decision factor.

There are many reasons for the current situation, and many institutional disincentives in place that prevent a move in the direction of more transparency into the true cost of operations. The reasons energy has not been systemically visible in the decision processes or incentive structure were logical in the past; energy was cheap and plentiful, and our operational missions assumed secure lines of communications making battlespace fuel delivery less of a burden. But things are different today.

The CNA and the DSB reports recommend that DoD take a series of specific actions that will begin moving it in the direction of well-informed and transparent decisions that affect energy security and have important national implications for national security and climate security. They can be distilled into three major themes.

First, we should incorporate climate change into our national security planning processes. This includes all levels of planning including our national intelligence assessments, national security strategy, national military strategy, quadrennial defense review, strategic planning guidance, requirements for future combat systems, and infrastructure investment plans. Scientific evidence that the climate is changing is unequivocal.

Second, we should put more emphasis on implementing two key policy memoranda of the past 2 years. One of these was issued by the Vice Chairman of the Joint Chiefs of Staff. It established an energy-related key performance parameter for the JCIDS process. The second was issued by the Under Secretary of Defense for Acquisition, Technology, and Logistics and established a policy to use the fully burdened cost of fuel for all acquisition programs. These two actions, well-implemented with sustained leadership support, could begin moving DoD toward fielding systems with greater endurance, better capability against asymmetrical threats, and based on emerging technologies that have important commercial applications and national benefits.

Third is to adopt a business case perspective when making investments that affect where DoD gets its energy and how it uses that energy. This applies to investment in both energy demand and supplies. Business case analyses are supposed to identify the investment opportunities that provide the greatest value to the entity. Here, the business case should identify the best value contribution to mitigate DoD's energy risks, considering all competing opportunities for those funds. One of the most fundamental choices for DoD funds is to invest in new energy supplies or demand reduction.

Summary.

Climate change, national security, and energy dependence are a related set of global challenges. Our energy sources both contribute to climate change and are at risk from climate change. Our security is threatened both by our high dependence on foreign sources of oil and by the climate change that high energy use helps bring about. To formulate effective climate mitigation and adaptation strategies, climate change must be a factor in our national security planning processes. If we look to unconventional fossil sources that increase carbon emissions to meet short-term fuel needs, it will be at the expense of investments in greater use of renewable sources and efficiency, putting us at greater long-term risk. DoD's reliance on high energy intensive operations compromises our military effectiveness in a number of ways. It creates operational vulnerabilities, jeopardizing mission success; and increases casualty rates, undermining popular support for the mission. It creates an unbalanced force structure, driving an unnecessarily large support "tail" at the expense of our operational "tooth." In some important ways, DoD's energy problem is like the nation's energy problem—we use too much. Demand side solutions to DoD's high fuel intensity can also help mitigate the nation's dependence on foreign oil and reduce future climate risks. As DoD deploys more energy efficient technologies in its combat systems, two important national benefits will result: Our industries will become more competitive in a global market that increasingly values efficiency; and national use of oil will decline as the technologies find their way into commercial products. Enacting the changes to DoD processes needed to achieve these benefits will require determined and sustained leadership. They require some changes in the factors DoD uses to make its most fundamental decisions affecting requirements, acquisition, force structure and funding priorities. It requires facing the realities of the true costs associated with high fuel use, including its drain on our operational effectiveness. These are factors DoD has not had to consider before, and is not currently equipped to consider. But the payoff for both DoD and the nation can be significant.

ENDNOTES - Morehouse

1. *National Security and the Threat of Climate Change*, Alexandria, VA: CNA Corporation, 2007.

2. *Ibid.*

3. Evidence that climate is changing is irrefutable, and comes from such direct observations as the thawing of permafrost; the lengthening of growing seasons; shifts in plant, insect, and animal ranges; and earlier flowering of trees. More recent terrestrial evidence shows that the rate of climate change is accelerating. Paleo-climate studies show atmospheric carbon is higher than at anytime in the past 650,000 years.

4. Previous speakers also noted that the climate is changing, and the concentration of atmospheric carbon is increasing. To what extent this is due to human input is debated. So too is how much change there will be, how rapidly these changes will occur, and where specific effects will take place. Debates include how

accurate are estimates of how much carbon and climate change can be attributed to anthropogenic sources, how much total change there will be in future, how fast it will occur, and the location of specific effects.

5. *National Security and the Threat of Climate Change*.

6. According to the Hubbert Peak theory, peak oil is the point or timeframe at which the maximum global petroleum production rate is reached. After this point, the rate of production will enter terminal decline along roughly the path of a bell-shaped curve.

7. These figures bear on the discussion earlier in the conference about equity. How do we come up with an equitable regime for dealing with carbon emissions and climate change?

8. *Defense News*, August 2006.

9. *More Capable Warfighting Through Reduced Fuel Burden*, Washington, DC: Defense Science Board, 2001.

10. General Paul Kern, Commanding General, Army Materiel Command, Tactical Wheeled Vehicles Conference, Monterey, 2002, quoted in *National Defense*, March 2002, p. 37.

11. Some other estimates: "Fuel Costs $13/Gal—Well to Tank—In Peacetime at Home," Army Research Lab brief to Defense Science Board, October 1999; "$25 at FEBA+100 km," Defense Science Board, 2001, p. 16; "Hundreds of Dollars [by air] . . . [600 km] Deep in the Battlespace," or "At least $40-50 If Overland," Defense Science Board, 2001, Executive Summary and p. 20; "$100-$600 In Theater Depending on 'Front Line' to 'Back Line' Separation in Distance, Terrain, Defense, Etc.," JASON, *Reducing DoD Fossil-Fuel Dependence*, 2006, p. 30; "$26 By Inflight Tanker," Defense Science Board, 2001.

Military Technology and Renewable Energy

David A. Sheets
Army Environmental Policy Institute

My name is Dave Sheets, and I am with the Army Environmental Policy Institute. We are a small policy group working under the Deputy Assistant Secretary of the Army for Environmental Safety and Occupational Health.[1] We work for Mr. Tad Davis. Those of you who are from North Carolina might know him as Colonel Davis. He was the Garrison Commander at Ft. Bragg a few years ago and actually started Ft. Bragg on the road to environmental sustainability. Therefore, he is pretty well-known in this part of the world. My purpose today is to focus on military technology and renewable energy. Time permitting, I will talk a bit about the Army Energy Program, discuss Army Energy and Sustainability Strategies, and outline some of the renewable energy technologies used by the Army.

The Energy Program.

As you know, the Army has an energy management program in place which aims to eliminate/reduce energy waste in existing facilities, increase energy efficiency, and improve energy security. There are a number of laws and regulations which set the parameters. Among them, I should mention the Clean Air Act (CAA) of 1990, which was the first major climate change program that affected the military. This banned the production and venting of Class 1 ozone depleting substances (ODS), created a Department of Defense (DoD) reserve, and allowed DoD use for Mission Critical Requirements (Crew Protection,

Fire Suppression, Tactical Vehicle A/C). A more recent document is EO 123423, "*Strengthening Federal Environmental, Energy and Transportation Management,*" January 2007.[2] This calls for, among other things, a 30 percent energy efficiency goal for federal agencies in 10 years, (50 percent more stringent than the Energy Policy Act [EPACT] 05). It also states that at least 50 percent of current renewable energy purchases must come from new renewable sources (in service after January 1, 1999); and it requires an increase of alternative fuels by at least 10 percent annually.

The Army does still have ODS in its legacy systems but has also met with some successes. Starting in base year 1992, it eliminated 80 percent of Halon 1301 use in Legacy systems, 9 percent of ODS solvents in industrial operations, and 98 percent of chlorofluorocarbon (CFC) use on Army installations.

We all know that the DoD is the largest single user in terms of U.S. energy consumption, but you may not necessarily know the details. In 2005, DoD spent $10.8 billion on fuel and consumed approximately 100M barrels of oil. That represents only 1.4 percent of U.S. use—Americans as a whole used about 2 million barrels per day that year—that is, 720 million barrels a year. Within the DoD, moreover, the Army is not the leader in regards to energy usage—it is the Air Force. In fiscal year 2005, the Air Force consumed 54 percent of the fuel, the Army 12 percent, and the Navy 33 percent.

What are the energy needs? Well, the Air Force flies a lot of jets and uses a lot of jet fuel. That is why they consume so much oil. What the Army does have is the largest number of utilities—35 percent of DoD utilities, as compared to 34 percent for the Air Force and 27 percent for the Navy. That is because, as you will see, we have a really large number of military installations, both within the United States and overseas. The Army Universe is big: We own 770 million square feet of buildings; we have 37,000 family housing units; we house 136,000 soldiers; and we own 13 million acres of land. We have about a million soldiers, which includes active duty, National Guard, and Reserves. The Army also has 209,000 civilians and 712,000 family members. In short, we have about 2 million people in the Army "family." That gives you some sense of who we are and where we are. Because of that, we do use a lot of energy. The energy used on installations is primarily not from oil, but rather from coal, natural gas, hydro-electric, and/or nuclear power.

Army Energy and Sustainability Strategies.

Now, what have we been doing to lean forward in the foxhole a little bit? There is the Army Strategy for the Environment which came out in 2004, and the Energy Strategy for Installations which came out in July 2005. There is also a new document that came from the Association of the United States Army (AUSA). Their February 2007 *Torchbearer Report* discusses Army Sustainability.

In regards to the Army Strategy for the Environment, I would like to stress that sustainability is the term that we use to tie us to the energy program. For many of you, it may be clear that the Army Environmental Program and the Energy Program should be joined at the hip. That is not necessarily the case. Environmental people used to be focused on compliance, restoration, and bugs and bunnies, not to mention the red cockaded woodpecker. (A few Ft. Bragg people know that woodpecker pretty well.) We didn't really have much input into the Army Energy Program, which was run through the Assistant Chief of Staff for Installation Management energy people.

Now, since the strategies are the "what we need to do," what about the "how will we do it?" For each of these documents, there is a strategic plan. The Army Energy and Water Campaign Plan for Installations came out last year. The Environmental Strategic Plan for implementing sustainability may be out this summer. The goal of *The Army Strategy for the Environment* is to foster a sustainability effort. Our triple bottom line is: mission, environment, and community. We have also added the economy, because we are looking to find ways to protect the environment that make economic sense. Sometimes when you take care of the environmental piece, you actually add value to the mission or the community, either by lowering costs, improving performance or adding some other benefits. One of the reasons we started with a green chemistry initiative is because we are looking for those win/win scenarios where we can actually lower waste and then make better quality products, while improving the environment.

What are the initiatives called for in *The Army Energy Strategy for Installations*? There are five plans outlined in this document:

1. To eliminate waste in existing facilities. With 770 million square feet of buildings, that is not an easy task.

2. To increase energy efficiency in renovation and new construction.

3. To reduce dependence on fossil fuels.

4. To conserve water resources.

5. To improve energy security.

Maybe you've heard of the Green Building Council's Leadership in Energy and Environmental Design (LEED) standards? There are silver, gold, and platinum certification levels. The Army is going for silver certification in all buildings beginning in 2008 and to do that, we will be reducing dependence on fossil fuels, conserving water resources, and improving energy security.

One of the big issues with the Army, at least where waste and efficiency are concerned, is accountability. This is a rather important issue, which I do not believe came up in yesterday's discussions. A lot of times the person working the light switch, constructing the building, or designing the vehicle does not pay the fuel bill once the structure is designed or built. That has been a problem for us, especially when we try to get that extra insulation put in the buildings, or to get people who do not pay for the utilities in their quarters to turn off light switches. That is something the Army is working on, and it is pretty straight forward. We should be designing green buildings, we should pay the extra couple of percent to get energy secure buildings. The issue is that often the people designing those buildings do not live with those buildings, so no emphasis is placed on doing those extra things to ensure efficiency and prevent waste.

Energy Technologies.

Now I am going to talk about some of these energy technologies. Of course, each of these topics deserves much more extensive treatment, but in the interests of time, I can only cover them briefly. Because many of you here are not Army people, I will start by telling you where we are located. The Army is located across the country, but we are not everywhere. The Army has a real presence in the Southeast and up the east coast. We are also across the Southwest and up in the lower Midwest. We really don't have a presence up in the Northwest except for Ft. Lewis, which is near Seattle, Washington.

Solar Power. First, I want to talk about solar technology. It has great potential for both a garrison and field environment. It is secure and reliable. It is also flexible to a wide range of loads. You can use solar batteries in your watch. At the same time, there is a 15 megawatt solar system being designed at Nellis Air Force Base on a 120 acre site near Las Vegas, Nevada. Solar systems are environmentally sustainable. There is no waste. You can connect with a grid. You can actually use solar panels in a place like Yuma Proving Ground, Arizona, where you may be 100 miles from a source of energy and be self-sustainable. If you hook them up to a wind unit, which works better at night, you can work the lights at all times.

Anywhere in the United States can benefit from solar power, even typically cloudy, rainy environments such as Ft. Lewis. For example, Germany has more solar panels than the United States, and their latitude is about even with Canada, so they get less sun intensity. Therefore, there is no real reason for us not to be able to do more.

This technology is being used by the Army. At Schofield Barracks in Hawaii, 5,000 homes will be powered by the sun. Instead of using the expensive crystalline solar panels, they will use the less efficient, but cheaper amorphous kind. They are going to roll these solar panels down the roofs. And they are going to generate about six megawatts of solar energy. Solar power is also being used in the field. A portable container sized unit, which has a little wind turbo, is on display up in Arlington, Virginia. It can be taken to the field to provide power to a small headquarters. The Army also has solar panels on tents, portable solar rucksacks that can be opened and laid flat to collect solar energy, and hand-portable battery chargers. These are really effective. The New Jersey National Guard is already using a 10 KW system on a roof. Therefore, the Army is making some good use of solar power.

Wind Power. Unfortunately, a lot of the big U.S. sources of wind power are out in the west, which only helps Ft. Lewis, Ft. Carson, Colorado, and a few other major bases. In the sand hills area of North Carolina and down into Texas and Louisiana, wind currently does not have much potential. However, as wind mills become more efficient, wind farms in even these lower speed areas might be economically feasible.

A recent report to Congress said that windmill farms hamper military readiness. It said that they affect radar and thus impede air defenses and possibly hamper training. Do not ask me to completely justify this. Though I am not an expert in any of these areas, I will say that we have some 13 million acres of land; there has to be room for some windmill farms somewhere in the western part of the country.[3]

Now, where can the Army use wind? The Army is looking to find ways to reduce the need for fuel on the battlefield. For example, the Rapid Equipping Force uses a trailer mounted wind unit. A National Guard unit based in Utah has a large wind turbine for power generation. The Army is looking into a way to combine solar panels and wind turbines. The solar panels provide energy during the day; and at night, when the wind works better, you use the wind.

Waste Energy. There is real potential in waste-to-energy systems for the Army. The Army can use waste energy, especially in both the field environment and in disaster relief. In the field, what are our wastes? Anybody that has ever seen soldiers in the battlefield knows the answer. There are piles of wood pallets and empty plastic water bottles that can be burned and used for energy. Almost 79 percent of the waste in the field has recoverable energy content. You can turn waste into electricity, heat, fuels, hydrogen, methane, and JP-8. There is a zero footprint base camp being developed where they try to recycle or reuse everything, and that includes waste water. Disasters also provide opportunities. If a hurricane knocks down trees, you typically burn them in a big bonfire, or you landfill them. That is energy that can be used in a waste-to-energy generator for possibly hooking up a small power grid and reestablishing electricity. This technology, too, is being pursued. We are working on a 50 kilowatt biomass generator, for example, in El Salvador.

Geo-Thermal Energy. There are two kinds of geo-thermal technologies, one where you get energy from the ground—highly usable—with thermal heat pumps, and these can be used anywhere. For example, Ft. Polk, LA, is now the world's largest geo-thermal installation—4,000 homes have been retrofitted there. The same thing could be done in lots of places.

Alternative Fuels and Fuel Cells. We are required, under law, to convert to 85 percent Ethanol blend (E-85) vehicles and 20 percent biodiesel (B-20) vehicles in our nontactical fleet. We are doing fairly well with this—48 percent of our light duty vehicles can use alternative fuels. So far, 12 installations have alternate fuel capacity. We have 23,000 vehicles, so the Army has, in fact, done a lot to get these alternate fuel vehicles. We also have a full range of fuel cell initiatives, from battery size to these generator-size tactical units. There are also fuel cell units for housing. Several units have been tested at Watervliet Arsenal in New York. The Army even has a fuel cell Segway that they are testing. Fuel cells have problems, however, especially in the battlefield. First, the platinum in a vehicle fuel cell costs about $3,000. You have to get the cost of platinum down—this is just too expensive. Second, the sulfur in JP-8 makes them ineffective by contaminating the platinum. Unless we bring methane to the battlefield, or find a way to use fuel cells that can handle the sulfur, fuel cells are going to be hard to use in the field.

The Army as a Leader.

One of the questions I was asked to address in my talk was whether the Army could or should be a leader in the effort to find a more sustainable future, and if so, how can the Army lead? One way it can do so is to develop new technologies and really use them. This is what we are doing. We have an extensive program working with fuel cells, batteries, and other new materials. I myself look at nanotechnologies. The Army Corps of Engineers is working on nano-enhanced mortar for buildings, which is pretty impressive. We are also looking at quantum dots and other materials. These, as you may have heard, are semi-conductors which can be used for increasing solar panels' efficiency.

Another thing the Army can do is to create a market for high efficiency devices. We have almost a billion square feet of buildings in the Army, if you add the overseas locations. We are engaged in sustainable installation efforts.[4] We can lead in developing heating and cooling initiatives that ensure high efficiency.

We are required under the new laws to use Energy Star certified equipment. Energy Star devices are only certified if they are within the top 25 percent for energy efficiency. Another way to reduce energy use is by using light emitting diode (LED) lights and the compact fluorescent lights (CFLs). Somebody yesterday asked what would happen if we switched to compact fluorescent lights? I have them in my house. You can get about six or eight 60 watt equivalent bulbs for about $10. Therefore, they are not that expensive, and they save you a lot of energy. Some of the naysayers right now say that there are problems with the mercury in CFLs, though the Environmental Protection Agency and Department of Energy dismiss this since the energy savings outweighs the mercury issue. My advice is to go out and buy the CFLs. If we take both individual and group conservation measures, we can do a lot of good, especially if this is done across the two million soldiers, civilians, and family members.

ENDNOTES - Sheets

1. For more about the Army Environmental Policy Institute, see *www.aepi.army.mil*.

2. Implementation Instructions were still in draft as of March 2007.

3. We should realize that even if we do get the farms, it won't help the Army much because this isn't where our footprint is.

4. See the Army Sustainability Program and the Army Energy Program.

Warfare and Climate Change

Karen Lesley Hulme
University of Essex, United Kingdom

Introduction.

At the strategic level, U.S. Field Manual (FM) 3-100.4 concerns environmental stewardship. The majority of the document details the ways in which the Army and Marine Corps must include environmental considerations in their military training, as well as in risk assessment and risk management. By contrast, it devotes a mere five pages to the rules governing combat operations. In this law of war section, at paragraph 4.3, the manual notes that "the public has been remarkably consistent in the last 25 years in its concern for global and local environmental degradation" and that "public" and "public pressure"/"public perception" have been an important factor in warfare. The manual continues, "Application of environmental protection in a given contingency will almost certainly differ from its application in the midst of close combat during a war."[1]

The manual contains an illustrative bell curve which demonstrates the level of environmental consideration appropriate to the various roles of the military—notably training, peacekeeping, and combat. At the top is training—where the level of environmental consideration is at its peak, and at the lowest point (the lowest level of environmental consideration) are combat operations. During armed conflict, environmental protection is never the priority. It is one of the lowest priorities. However, that does not mean that it is completely absent. The laws of war have addressed environmental protection since the 1970s, and, although the laws do not refer directly to climate considerations or climate change, there is much valuable mileage in the existing environmental considerations in wartime.

The topic of warfare and climate change raises a number of questions:

- Armed conflict is inherently destructive of the environment, but how destructive is it?
- Are there any legal limits on the type or scale of environmental damage that the armed forces are allowed to cause during armed conflict?
- When it comes to the issue of environmental protection, do the laws of war coincide with climate change abatement strategies? And what, practically, can the armed forces do in war to reduce their carbon footprint?

To answer these questions, this section will analyze the laws that afford protection to the environment in peacetime—in those times outside combat. The phrase "peacetime" here refers to international environmental laws. The central question is whether these peacetime laws continue to apply during armed conflict? I will then go on to analyze the laws of armed conflict which specifically apply during combat operations.

Peacetime Environmental Laws *(Jus Pacis)*.

In 1979 states recognized the need for global treaty rules to address the pollution of the atmosphere. Although states had adopted a number of international treaties in the 19th and the early part of the 20th centuries, it was in 1972 that international law truly embraced environmental protection more broadly.[2]

In 1972 states met in Stockholm under the auspices of the United Nations (UN) for the UN Conference on the Human Environment. At Stockholm, states adopted a set of principles in the form of a Declaration,[3] essentially a nonbinding instrument which would lay the foundations for environmental protection for the future. According to Principle 2, "the natural resources of the earth, including the *air*, water, land, flora and fauna and especially representative samples of natural ecosystems, must be safeguarded for the benefit of present and future generations through careful planning or management, as appropriate" (emphasis added).

The Stockholm Declaration introduced the notion of safeguarding such natural resources as the air, for the benefit of mankind—including future generations. Then in 1979 states agreed to the first global measure to tackle the problem of air pollution with the Convention on Long-Range Transboundary Air Pollution.[4] This treaty was drafted to tackle the transboundary problem of acid rain, caused by industrial and vehicular emissions of sulphur dioxide. States agreed to limit their emissions of sulphur dioxide,[5] and in later Protocols nitrogen oxides,[6] volatile organic compounds (VOCs),[7] heavy metals,[8] and ammonia.[9] The next treaty governing atmospheric pollution was the 1985 Vienna Convention for the Protection of the Ozone Layer,[10] which established emissions limits on CFCs and other ozone-depleting compounds.

To tackle the issue of global warming, states adopted the 1992 United Nations Framework Convention on Climate Change,[11] and the 1997 Kyoto Protocol,[12] the latter of which introduces timetabled emission reductions of greenhouse gases for party states including carbon dioxide, methane, and nitrous oxides.[13] The United States has so far refused to ratify the Kyoto Protocol. Therefore, while the United States is bound by the more vague 1992 treaty obligations to limit emissions and adopt policies to mitigate climate change, it is not bound by the specific timetabled reductions established in the Kyoto Protocol.

The essential question is whether these rules of peacetime (*jus pacis*) environmental protection continue to apply in situations of armed conflict. According to state practice, the answer for two or more belligerent states *inter se* is "probably not." The issue generally comes down to the compatibility of the environmental obligations with a situation of armed conflict. And so peacetime environmental obligations contained within a specific treaty may continue to apply only so far as those obligations are compatible with the situation of armed conflict. As a result, obligations to reduce certain harmful emissions might be deemed to be generally incompatible with a situation of armed conflict. Consequently, such incompatible obligations would no longer be applicable (to the extent of the incompatibility) during conflict between belligerent states *inter se*. A belligerent state that is party to Kyoto, however, still has to achieve its emissions limits at home. However, it is not obliged to adhere to these requirements on the battlefield. Environmental obligations between a belligerent state and nonbelligerent state, on the other hand, do generally continue to apply, as the situation is akin to that of "peacetime."

The Laws of Armed Conflict *(Jus in Bello)*.

The laws can be arranged into two categories: laws providing ancillary environmental protection; and laws designed to protect the environment.

Laws Providing Ancillary Environmental Protection. Since 1977, the laws of war or humanitarian law have afforded protection to the natural environment.[14] Generally, this body of rules establishes what the military forces can and cannot do in situations of armed conflict. Primarily, therefore, the laws of war establish rules for lawful weapons and targeting. At the core of humanitarian law is the principle that means and methods of warfare are not unlimited.[15] This means essentially that the military can only use lawful weapons and tactics. When it comes to the environment, of course, it will generally benefit from the prohibition of most types of weapons, such as chemical weapons, land mines, herbicides, and other biological weapons.

Secondly, the law establishes the principle of distinction; [16] this is the principle that the military can only target military objectives, not political or economic objectives, although some states do include economic targets within a broader definition of military objectives. As regards the environment, it is *prima facie* a civilian object, and so it cannot be subject to attack. However, the environment may be attacked where it fulfills the requirements of a "military objective,"[17] such as when the enemy uses forest or plants as concealment, as in Vietnam. So although not primarily a "military objective," the enemy's military use of it changes its status into a military objective.

The third principle is that of proportionality,[18] which concerns the notion that when attacking military objectives, any damage caused to civilians and civilian objects cannot be excessive to the military advantage anticipated to be gained from the attack. This is usually termed collateral damage. Since the environment is *prima facie* a civilian object, any collateral damage caused to the environment must be included in the quantification of collateral harm and so cannot be excessive. However, it is very difficult to measure the two parts in the balance, and the level of collateral environmental damage that can be caused before a particular attack would be deemed disproportionate is probably quite high.

The fourth principle is that of military necessity, and leads to the fundamental prohibition on the wanton destruction of property,[19] that is, destruction without a military purpose. This prohibition clearly benefits the environment, since the military cannot wantonly destroy the environment. Finally, since a broad interpretation of the environment might include crops and agriculture, Article 54 of Protocol I specifically prohibits the attack or destruction of "objects indispensable to the civilian population," which includes foodstuffs.

In summary, the fundamental principles of the laws of armed conflict stipulate:

- Means of warfare are not unlimited—use only lawful weapons/tactics.
- Distinction—Military vs. civilian—only military objectives may be attacked.
- Proportionality—attack of military objectives must not cause excessive collateral (civilian) harm.
- Military Necessity—prohibition on wanton destruction of property (i.e., without military purpose).

Article 54 of Protocol I stipulates (in part):

Protection of objects indispensable to the survival of the civilian population:

> 1. Starvation of civilians as a method of warfare is prohibited.
>
> 2. It is prohibited to attack, destroy, remove or render useless objects indispensable to the survival of the civilian population, such as food-stuffs, agricultural areas for the production of food-stuffs, crops, livestock, drinking water installations and supplies and irrigation works, for the specific purpose of denying them for their sustenance value to the civilian population or to the adverse Party, whatever the motive, whether in order to starve out civilians, to cause them to move away, or for any other motive.
>
> 3. The prohibitions in paragraph 2 shall not apply to such of the objects covered by it as are used by an adverse Party:
>
> (a) as sustenance solely for the members of its armed forces; or
>
> (b) if not as sustenance, then in direct support of military action, provided, however, that in no event shall actions against these objects be taken which may be expected to leave the civilian population with such inadequate food or water as to cause its starvation or force its movement.

Laws Designed to Protect the Environment. How do the laws of armed conflict protect the environment? Environmental protection in wartime was adopted following U.S. tactics in Vietnam. The United States caused the destruction of three-quarters of a million acres of land in Vietnam with the use of Rome plows, and wide scale damage to crops and mangrove with the use of herbicides. So in 1977, with global environmental awareness heightened by the 1972 Stockholm Declaration and U.S. tactics in Vietnam, states adopted Additional Protocol I to the four 1949 Geneva Conventions. And with the adoption of Protocol I, states included in the laws of armed conflict two provisions which specifically applied to the environment. The first provision was Article 35(3) which stipulates, "It is prohibited to employ methods or means of warfare which are intended, or may be expected, to cause widespread, long-term and severe damage to the natural environment."

The second provision is Article 55 which builds upon this provision in three important ways: (1) the addition of an obligation of "care," (2) the linking of the prohibition specifically to human harm, and (3) a prohibition on reprisals against the natural environment. It is the first of these three aspects which may best improve environmental protection during armed conflict—the general obligation on states to take care to protect the environment against widespread, long-term, and severe damage. Therefore, Protocol I includes a specific prohibition on the use of weapons and tactics ("means and methods of warfare") which may forseeably cause a serious level of environmental damage. The element of foreseeability dictates that states have to stop and ponder the possibility that a particular weapon or tactic might breach the threshold of harm. Finally, as regards the development of new weapons, states must ensure that their effects will conform to the rules and principles of humanitarian law binding on that state. This obligation is enshrined in Article 36 of Protocol I and derives from U.S. practice.

The two provisions, Articles 35(3) and 55, are drafted so as to protect the natural environment against harm—the threshold of which is actually set very high (widespread, long-term, and severe). Yet the protections have an anthropocentric basis—the idea that any environmental damage will have knock-on effects on people.

In 1977 the United States and Russia also adopted a second treaty concerning tactics of warfare which might have environmental effects, the United Nations

Convention on the Prohibition of Military or Any Other Hostile Use of Environmental Modification Techniques (ENMOD).[20] It was adopted in New York and currently boasts some 72 party states, including the two main negotiating states, Russia and the United States, and China since 2005. ENMOD came about as a result of the global fear evoked by U.S. cloud seeding techniques in Vietnam, tactics which appear to have been in design since the 1940s; and the panic caused by scenarios that appear more at home in science fiction novels than international law treaties. Cloud seeding with lead or silver iodide was attempted by the United States in Vietnam to flood the Ho Chi Minh trail, with the hope of cutting off the supply route from North Vietnam down to the South.

The Convention establishes a prohibition on military or any hostile use of ENMOD techniques which cause destruction or injury to the enemy. Accordingly, Article I of the Convention stipulates, "Each State Party to this Convention undertakes not to engage in military or any other hostile use of environmental modification techniques having widespread, long-lasting, or severe effects as the means of destruction, damage, or injury to any other State Party." Thus what is prohibited is the use of such "environmental modification techniques" (EMT)—manipulating the environment in some way—to cause harm to another party state, for example the creation of a tidal wave to destroy enemy fortifications in another party state, or the use of herbicides to upset the ecological balance and cause harm to human health.[21] The Convention defines "environmental modification techniques" at Article II as "Any technique for changing—through the **deliberate** manipulation of natural processes—the dynamics, composition or structure of the Earth, including its biota, lithosphere, hydrosphere, and atmosphere, or of outer space" (emphasis added). And finally, it is a war crime to intentionally launch an attack in the knowledge that such an attack will cause widespread, long-term, and severe damage to the environment, which is also "clearly excessive in relation to the concrete and direct overall military advantage anticipated."[22] Unlike the prohibition in Protocol I, for individual criminal liability, the environmental damage has to be intentional.

Applying the Laws of Armed Conflict.

How does climate change affect the military on the battlefield? The issue under consideration is whether the protection of the environment (and climate change) creates obstacles for the military or opportunities.

The Military's Carbon Footprint in War. The first issue is the military's carbon footprint in war. One of the greatest contributors to climate change is the burning of fossil fuels. Other papers have outlined the measures that the U.S. military is taking to reduce the military's impact on climate change as it installs wind and solar power on its bases and training facilities, and experiments with the use of bio-fuels. It is these kinds of activities which will lead to a reduction in the emissions levels of greenhouse gases by the military, because there is absolutely no way that a military commander would ground his troops for the day because he thinks his carbon footprint is too high.

Bombs and other explosive weapons produce carbon dioxide emissions on detonation, and the bigger the bomb, generally the bigger the carbon footprint. Do the laws of war address this issue at all? Yes, to some degree, they do. The principle of economy of force is relevant. This principle is a pragmatic one and essentially means "why waste money and resources using 10 bombs when one will do the job," or "why waste money and resources using a larger bomb when a smaller one will do the job." Also pertinent is the principle of proportionality which is concerned with reducing nonmilitary (or civilian) harm. Since a bigger bomb is more likely to inflict excessive harm on the civilian population than a smaller bomb, this principle also encourages the use of smaller bombs. As weapons are destructive of the environment, bombs should logically, perhaps, be phased out. However, can this be done? In fact, the law forbids the use and development of weapons that do not comply with the laws of wars but goes no further. Unless and until more specific rules are drafted which outlaw those weapons, they will remain legal. More to the point, though, the armed forces will not give up weapons that are lawful, if they are also valuable.

Furthermore, if the armed forces were forced to give up carbon emitting bombs, this would actually lead to a more controversial issue, in that this would then suggest greater use of noncarbon-emitting weapons such as depleted uranium. This would be very similar to peacetime arguments in favor of the use of nuclear power instead of fossil fuel.

A third issue in this category is the targeting of installations that emit greenhouse gases. Some installations, when bombed or damaged, might release dangerous gases or leak dangerous chemicals—including greenhouse gases. Do the laws of war prohibit the military from targeting such installations? If one starts with basic principles of the laws of war, the answer is that many such facilities will not fulfill the definition of a military objective. If an object cannot be classified as a military objective, then it cannot be attacked. While factories producing goods used by the military—and oil refineries, for example, have been classified as military objectives—U.S. policies during the 1991 Gulf Conflict are slightly promising regarding environmental protection. Probably the most memorable image of the Gulf Conflict is the 590 burning oil wells in Kuwait, destroyed and set alight by Iraqi troops.

In the opinion of the United States, these acts were contrary to the laws of armed conflict due to the wanton and unnecessary nature of their destruction.[23] While this opinion may have been due in part to the U.S. appreciation of the environmentally-destructive nature of the burning wells, the greater consideration undoubtedly came down to the lack of military utility of the destruction. This assessment was clearly influenced by the fact that the United States was being supplied with oil by Saudi Arabia. That meant that the Iraqis could not argue that they destroyed the oil fields in order to deny them to the U.S. forces—in other words, out of military necessity. A positive aspect, therefore, from the environmental perspective (and climate change in particular) is that the United States made a targeting distinction during the Gulf Conflict not to target crude oil installations. While U.S. forces were willing to target oil refineries which would produce a finished, usable product (petrol), they decided not to target the long-term mining of crude oil, and so, in effect, crude oil installations did not then routinely form military objectives.[24]

Targeting Carbon Sinks. The rainforests and other forested areas around the world provide a valuable sink for carbon dioxide emissions, and the possibility of the sequestration of carbon emissions in these sinks can counterbalance the high level of greenhouse gases in the environment. Would a state's armed forces, therefore, be able to destroy carbon sinks, such as forests or other kinds of plant cover, legitimately? Furthermore, would a state's armed forces be able to legitimately destroy agriculture, so as to increase or hasten the release of methane into the atmosphere and carbon dioxide into the soil? The laws of war already prohibit the targeting of anything that does not have a military advantage. And so the principle of distinction would, first and foremost, prevent the direct targeting of forests and crops, for example, that did not serve a military purpose. In modern international warfare, it is rare for a state to adopt tactics of crop destruction so as to deny sustenance for the enemy. Any attack on crops is more than likely to be a result of use by the enemy of the crops as cover or concealment, which would bring the crops within the definition of military objective and hence open for attack. If sustenance were the issue, of course, the military would need to refer to Article 54 of Protocol I which specifically outlines the additional steps in the procedure for such objects to be attacked.

In addition, if the forest were to fulfill the definition of a military objective, the environmental protection included within Article 35(3) would remain applicable to ratifying states, and provides an absolute threshold of harm which could be caused to the environment. The general view in international law of this provision, however, is that the threshold, albeit absolute, is also set very high and is possibly too high to have any impact on conventional warfare (that is outside the realms of nuclear, chemical, and biological warfare). One of the most important ways, however, of influencing the military on this point, might be to encourage it to view any attack on a carbon sink (for example the destruction of a forest), even if militarily necessary, as causing severe environmental damage. Hence if the destruction of a carbon sink comes to be viewed routinely as falling within the threshold of "severe" harm, the military may be more inclined to avoid such tactics. In effect, greater protection of carbon sinks could be a question of the interpretation of an existing threshold, as opposed to the negotiation of a new instrument.

Climate Change as a Weapon of War.

Finally, the question was raised as to the whether climate change, or the effects or causes of climate change, can be used as a weapon of war. If and when climate change raises the sea level, can the armed forces attack sea defenses with a view to flooding their enemy and killing their enemy in that way? Such tactics have been used throughout history. Such tactics were recently used by China in the second Sino-Japanese war, when China dynamited one of the dikes of the Yellow River in order to prevent the onslaught of Japanese troops. In using this tactic, the Chinese killed thousands of enemy troops but also killed hundreds of thousands of their own people in the process.[25] Today, such tactics are prohibited by Article 56 of Protocol I, but only to the extent that the resulting floods would cause severe (and excessive) civilian losses.

Article 56 of Protocol I stipulates (in part);

> Protection of works and installations containing dangerous forces:
>
> 1. Works or installations containing dangerous forces, namely dams, dykes, and nuclear electrical generating stations, shall not be made the object of attack, even where these objects are military objectives, if such attack

> may cause the release of dangerous forces and consequent severe losses among the civilian population. Other military objectives located at or in the vicinity of these works or installations shall not be made the object of attack if such attack may cause the release of dangerous forces from the works or installations and consequent severe losses among the civilian population.
>
> 2. The special protection against attack provided by paragraph 1 shall cease:
>
> (a) for a dam or a dyke only if it is used for other than its normal function and in regular, significant, and direct support of military operations, and if such attack is the only feasible way to terminate such support . . .
>
> 3. In all cases, the civilian population and individual civilians shall remain entitled to all the protection accorded them by international law, including the protection of the precautionary measures provided for in Article 57. If the protection ceases and any of the works, installations, or military objectives mentioned in paragraph 1 is attacked, all practical precautions shall be taken to avoid the release of the dangerous forces.

Beyond specific protections such as this, can the armed forces try to accelerate or induce climate change? In other words, would it be legitimate for the military to use the effects of climate change as a tool or weapon so as to destroy the enemy? For example, can the military deliberately try to melt the polar ice cap, cause flooding of enemy territory, cause drought or induce famine, create a hole or a "window" in the ozone layer above enemy territory to burn the enemy—as legitimate weapons of warfare? Such tactics may sound like science fiction, but these issues were particularly prevalent in the 1970s following the use of cloud-seeding (rain-making) tactics by U.S. forces in the Vietnam conflict.

During Operation POPEYE in Vietnam, the United States used cloud seeding techniques to induce rainfall and so flood the Ho-Chi-Minh Trail. Apparently, there were some 2,600 flights involving some 47,000 units of cloud-seeding material. And such flights appear to have occurred outside Vietnam in Cambodia and Laos in breach of international law. In addition to trials in wartime situations, many countries have tried weather modification techniques for peacetime uses, too. However, military or hostile uses of such weather modifications have been prohibited in the 1977 ENMOD.[26]

The ENMOD Convention stipulates:

> Article II
>
> 1. Each State Party to this Convention undertakes not to engage in military or any other hostile use of environmental modification techniques having widespread, long-lasting or severe effects as the means of destruction, damage or injury to any other State Party.
>
> 2. Each State Party to this Convention undertakes not to assist, encourage or induce any State, group of States or international organization to engage in activities contrary to the provisions of paragraph 1 of this article.
>
> Article II
>
> As used in Article I, the term "environmental modification techniques" refers to any technique for changing—through the deliberate manipulation of natural processes—the dynamics, composition or structure of the Earth, including its biota, lithosphere, hydrosphere, and atmosphere, or of outer space.
>
> Article III
>
> 1. The provisions of this Convention shall not hinder the use of environmental modification techniques for peaceful purposes and shall be without prejudice to the generally recognized principles and applicable rules of international law concerning such use.

The Convention does not prevent peacetime uses and the peacetime development of such environmental modification techniques (see Article III). And so a number of states have developed weather modification programs, including the dispersal of rain from clouds already formed, the creation of rain using chemicals drawing water up from lakes (precipitation enhancement), the creation or dispersal of fog, and the creation of storms, possibly including lightning. Clearly, successful peacetime uses would lead to suggestions of wartime application. So, would states abandon the ENMOD Convention if such weapons became viable? This is a difficult question to answer, but if the ENMOD Convention were abandoned, it could open the door to some very deadly new weapons, and ones that would appear to fly in the face of a responsible approach to climate change. Wartime applications could include the creation of both rain or storms to enhance the abilities of a military's own forces and friendly forces, and to degrade the abilities of the opposing forces. A good example of such military applications would be the creation of rain on demand, so as to flood enemy territory and hinder its operations, or to burst rain clouds earlier so as to keep areas dry for its own forces—and hence enhance its own environmental conditions.

The ENMOD Convention only prohibits environmental modification techniques which have effects above a particular threshold, and while the threshold of harm is relatively low, it is not an absolute prohibition even in wartime on weather modification techniques. The threshold incorporated into the ENMOD convention is "widespread, long-lasting, or severe" effects. Thus only environmental modification techniques with widespread, long-lasting, or severe effects—that cause destruction, damage, or injury to the enemy—will fall within the scope of the Convention. In an Understanding attached to the 1977 Convention, the three terms were understood to mean: "widespread—an area of several hundred square kilometres"; "long-lasting—several months or more, or approximately a season"; and "severe—severe or significant disruption or harm to human life, natural or economic resources, or other assets."[27] As opposed to the cumulative threshold of harm contained in Article 35(3) of Protocol I, the three criteria in ENMOD are disjunctive: only one of which need be fulfilled to constitute a breach of the Convention. Thus a small-scale precipitation enhancement operation may be lawful. Furthermore, a reading of the Convention appears to allow the use of such environmental modification techniques in defense, where the enemy (any other state party) will not be affected by damage or injury, for example, where the defending state creates drier conditions to enable its forces to maneuver or attack.

Conclusion.

The purpose of this section was to consider the impact of the military on the environment and the environment on the military. It was also to consider what could be done to reduce the military's carbon footprint. From a wartime perspective, are the armed forces hampered by environmental restrictions? The military is undoubtedly a massive polluter, particularly when it comes to the emission of greenhouse gases, and the current legal restrictions have little impact on these sources of pollution. More is likely to be achieved by the adoption of clean technology by the military than in pushing for regulation of fossil fuel emissions in wartime. As regards targeting, on this point the surprising conclusion is that climate change mitigation and the laws of armed conflict coincide to a greater extent than would be first thought. While the environment is an unfortunate victim of war, it should rarely be a direct target, and regulations adopted in the aftermath of the Vietnam conflict may serve to rule out many exploitative uses of an environment affected by climate change (including all of the new rules in Protocol I and ENMOD).

ENDNOTES - Hulme

1. *Environmental Considerations in Military Operations*, Field Manual (FM) 3-100.4/MCRP 4-11B, June 15, 2000, available at *www.globalsecurity.org/military/library/policy/army/fm/3-100-4/index.html*, accessed June 23, 2007.

2. For general environmental law texts, see Philippe Sands, *Principles of International Environmental Law*, Cambridge, UK: Cambridge University Press, 2003.

3. 1972 United Nations Conference on the Human Environment, Stockholm, Sweden, YUN, Vol. 26, 1972, p. 319.

4. 1979 Convention on Long-Range Transboundary Air Pollution, Geneva, Switzerland, entered into force March 16, 1983, *International Legal Materials* (ILM), Vol. 18, 1979, p. 1442. Note that *International Legal Materials* is a useful tool for those trying to find the raw material of international law which can be hard to find.

5. 1988 Protocol on the Reduction of Certain Emissions or their Transboundary Fluxes by at Least 30 Per Cent, Helsinki, Finland, entered into force September 2, 1987, *ILM*, Vol. 27, 1987, p. 707. See also the 1994 Protocol on the Further Reduction of Sulphur Emissions, Oslo, Norway, entered into force August 5, 1998, *ILM*, Vol. 33, 1994, p. 1540.

6. 1988 Protocol Concerning the Control of Emissions of Nitrogen Oxides or Their Transboundary Fluxes, Sofia, Bulgaria, entered into force February 14, 1991, *ILM*, Vol. 28, 1988, p. 214.

7. 1991 Protocol Concerning the Control of Emissions of Volatile Organic Compounds or Their Transboundary Fluxes, Geneva, Switzerland, entered into force September 29, 1997, *ILM*, Vol. 31, 1992, p. 568.

8. 1998 Protocol on Heavy Metals, Aarhus, Denmark, entered into force 29 December 2003, available at *www.unece.org/env/lrtap/full%20text/1998.Heavy.Metals.e.pdf*, accessed July 20, 2003.

9. 1999 Protocol to Abate Acidification, Eutrophication and Ground-level Ozone, entered into force May 17, 2005, available at *www.unece.org/env/lrtap/full%20text/1999%20Multi.E.Amended.2005.pdf*, accessed July 2, 2007.

10. 1985 Convention for the Protection of the Ozone Layer, Vienna, Austria, entered into force September 22, 1988, *ILM*, Vol. 26, 1987, p. 1529.

11. 1992 United Nations Framework Convention on Climate Change, New York, entered into force March 24, 1994, *ILM*, Vol. 31, 1992, p. 849. There are currently 191 party states.

12. 1997 Protocol to the United Nations Framework Convention on Climate Change, Kyoto, Japan, entered into force February 16, 2005; ILM, Vol. 37, 1998, p. 22. There are currently 175 party states.

13. Others include hydrofluorocarbons, perfluorocarbons, and sulphur hexafluoride.

14. See Article 35(3) and 55 of the 1977 Protocol (I) Additional to the Geneva Conventions of 12 August 1949, and Relating to the Protection of Victims of International Armed Conflicts, Geneva, Switzerland, entered into force December 7, 1978; *ILM*, Vol. 16, 1977, pp. 1391-1441.

15. This principle was codified at Article 35(1) of Protocol I.

16. Article 48, Protocol I.

17. Article 52(2), Protocol I.

18. Article 51(5)(b), Protocol I.

19. Article 3(b) of the 1993 Statute of the International Tribunal for the Former Yugoslavia, May 25, 1993, as amended May 13, 1998, *ILM*, Vol. 32, 1993, p. 1159; and Article 8(2)(a)(iv) of the 1998 Statute of the International Criminal Court, which refers to "extensive destruction."

20. Entered into force October 5, 1978. *ILM*, Vol. 16, 1977, pp. 88-94.

21. A prohibition on the use of herbicides in this way was included in the 1992 General Assembly Resolution 47/52, entitled "General and Complete Disarmament," A/RES/47/52, 81st plenary meeting, December 9, 1992.

22. Article 8(2)(b)(iv), 1998 Statute of the International Criminal Court, entered into force July 1, 2002, ILM, Vol. 37, 1998, p. 999. The United States is not party to the ICC Statute, and it is difficult to establish the precise status in customary law of this provision.

23. *Conduct of the Persian Gulf War*, Final Report to Congress Pursuant to Title V of the Persian Gulf Conflict Supplemental Authorization and Personnel Benefits Act of 1991, Public Law 102-25, April 1992, Washington, DC: Department of Defense, 1992, pp. 610, 623, available at *www.ndu.edu/library/epubs/cpgw.pdf*, accessed July 30, 2007.

24. *Ibid.*, p. 99.

25. See Margaret T. Okorodudu-Fubara, "Oil in the Persian Gulf War: Legal Appraisal of an Environmental Warfare," *St. Mary's Law Journal*, Vol. 23, 1991, p. 152.

26. Entered into force October 5, 1978. *ILM*, Vol. 16, 1977, pp. 88-94.

27. See, generally, Paul Fauteux, "The Gulf War, The ENMOD Convention and the Review Conference," United Nations Institute for Disarmament Research Newsletter, No. 18, 1992, p. 6.

Commentator

Christopher H. Schroeder
Duke University

Just on the last point about the contribution of the military to this problem, the figures I have from a United Nations (UN) report are that global military operations in 2003 contributed about 5 percent of global greenhouse gasses. That is about the size of Japan, which is the fifth largest national contributor. So you can look at global military operations as about the equivalent of Japan.

You have just heard three very interesting papers on addressing the question, "Dealing with Climate Change: Opportunities or Obstacles for the Military?" I think you could just as well flip the question and ask, "Dealing with the Military: Opportunities or Obstacles for Climate Change?" The papers would read just as well because we have been answering both the question of "To what degree is the military going forward and contributing to the problem of climate change?" and "How might we do something about it? How might we turn that relationship around?"

This panel was framed around the question of whether or not climate change posed obstacles for the military or presented it with an opportunity. Of course, as Doug anticipated, the answer to that either/or question is, "Yes" and "Yes." It is both of those things. And it is interesting to think a little bit about what variable affects the flow of that relationship—what variable makes it a positive or a negative relationship between the two. I think one dimension is the time horizon. The more short-term you think about this problem, the more the military and climate change are in conflict; and the more long-term your thinking is, the more there are opportunities for the military, in fact, to become part of the solution.

The short-term was illustrated by Karen's talk. The shortest-term thinking in the military, and for the country, in thinking about the use of the military, is when we are in a conflict. That gets your attention. There is nothing more arresting than feeling the country is at risk, as we experienced after September 11, 2001 (9/11). And, of course, what really affected attitudes after 9/11 wasn't the tragedy itself. It was worrying about that second attack. And the military mobilized. We invaded Afghanistan and then Iraq.

As Secretary Donald Rumsfeld said in one of his press conferences, "You go to war with the Army that you have." So we will go to war with the Army that we have, and climate change be damned. You are not going to get the military to change the way they conduct operations in time of hostility just because you are worried about climate change. You can, however, structure your procurement, plan, and modify your logistical thinking in anticipation of that event. And I think the two other presentations brought that point out very well.

I have taught environmental law since before there was such a subject. At one time, I could have summed up the relationship between the armed services and environmental law by saying that the military has never seen an environmental control or regulation for which it would not wish to have an exemption. You saw that in the very early days of the Clean Air Act and the Superfund, the Clean Water Act, and so on.

Greenhouse gas is no exception. We sent our negotiating team in 1997 to Kyoto, Japan, with instructions to negotiate a national security exemption from whatever caps we ended up agreeing to. Of course, we did not agree to any, and the convention did adopt a much narrower exemption for armed conflict that has UN approval, but that wasn't what the military was seeking. It was actually seeking a national security exemption, and it has sought that in other situations as well. The military often asserts that anything that affects preparedness is a national security concern. So if you impose any kind of environmental constraint on me (ie., the military) that may limit the degree of flexibility that I have to structure my training, my personnel activities, my procurement, or whatever else it might be, there ought to be an exemption from that as well.

This is often the initial reaction. However, as these presentations illustrated with respect to the greenhouse gas problem, over time attitudes can shift. The external concern becomes internalized in the organization, and then you begin to see convergence between the civilian priorities and military priorities on environmental issues.

There is an interim period that follows the period when the military is in its resistant mode, but before it has internalized a problem. During this phase, the military still has to deal with a lot of issues. A lot of these have to do with base closings and the problems of hazardous waste around military sites. But I hazard to say that the military is coping with those problems as they build new installations, and that they are more responsible now than they used to be about what kind of footprints they leave.

The two talks by David and Tom give us some kind of advanced insight into current thinking within the defense establishment. There appears to be a realization that the military is uniquely situated in some ways to do things which might help solve the greenhouse gas problem. We have reason to be hopeful here, because it is, to a certain degree, in the military's self-interest to think about such problems. As was noted, our troops in Iraq **want** to reduce the number of convoys that have to travel just to bring petroleum products into forward installations.

An article in *The Wall Street Journal* written at the start of this year reinforces this point. The authors wrote, "The Pentagon's planning to spend more than $2 billion in the next 5 years on energy initiatives, which could help spur development of energy sources for use in other sectors." Then they quoted Marine Corps Major General Richard Zilmer, commander of the U.S. forces in Anbar Province, who was asking for a shipment of mobile solar panels and wind turbans to supplement gas-guzzling generators. He said, "The military's dependence on fuel for power generation could reduce the number of road-bound convoys and U.S. casualties."

The Army's rapid equipment force is trying to arrange for delivery to the general of hybrid power stations and solar panels. This is a high priority and is being done out of military self-interest. But it will have another consequence—one noted, I think, by both David and Tom. It means that the power of the military procurement process will be brought to bear on the climate-change problem. It will create new markets. A lot of the improvements we made with respect to environment and energy and efficiency issues have been driven in the past 30 years by entrepreneurs who are trying to develop new technologies. There is nothing to help stimulate a new technology more than having a pretty firm expectation that there will be a market for that technology.

Government procurement generally, and military procurement specifically, is an area in which American domestic policy could make a big difference. This can be seen in the 2005 Energy Policy Act. The civilian leaderships here decided that fleet purchases in the military will reflect a certain percentage of E85-capable vehicles.

Energy efficiency rating requirements may also prove useful. It is true that it can be a problem when government gets in the business of picking winners and losers in a time of rapid technological change. We could screw things up by riding the wrong horse or developing policies that are too brittle and do not have the capacity to reflect the best innovations that are going on. People have always worried about this. However, there are ways to mitigate those concerns. And there is an urgent need to stimulate the kind of technological improvements that will more rapidly ramp us off the carbon-based energy economy and on to alternative systems.

We have so many promising technologies that are cost-ineffective today. They need one, two, or three technological changes that people have theorized might be capable, but the changes have not yet been achieved on the bench, and certainly they have not been brought up to commercialization. Having proactive procurement policies is a way to stimulate getting more people into that business and accelerating the transition.

So I thank you for the opportunity to discuss three very interesting topics. I want to leave you with a question. Planning is great. But as we heard already this morning and earlier, plans do not **do** anything. Having people in the room who are sophisticated about environmental problems or energy problems is another step in the right direction. But you have also got to have the ear of the leadership. And the leadership must drive the change in all kinds of mundane ways, like structuring of the DoD procurement manager's incentives or making sure the priorities are played out in the specs that are actually issued. Otherwise all these plans just sit inside of volumes that look very elegant, but do not have traction on the ground.

That has always been the big problem in trying to move a large-scale institution even a couple of deflections away from the way it is progressing. So any insights as to how you translate the intimations of progress into even more substantial progress, would be something I think would be beneficial to discuss some more. So thank you very much.

Discussion

David Sheets: The one thing I did not discuss was metering requirements. Since this is not an energy group, you may not be aware of the Energy Policy Act of 2005, or, in fact, Executive Order 13423. These basically strengthen the government through energy savings—they call for metering on all the buildings. This means that we will be able to look at some 770 million square feet of buildings and find out what people are using by way of energy. We can start identifying trends within government quarters. Military families are allowed to use a certain amount of electricity, and after that, they have to start paying for their own utilities. That means we are being held accountable in a way we were not in the past. When I was a lieutenant and had a chance of getting an air conditioner, I did not care whether it was 220 volts or 110 volts. I was not paying for the electricity, so I got whatever was cheapest. That is not going to happen so much any more because people are now required to care. So we are not becoming more energy efficient simply because we want to: there are, indeed, some national policies that are forcing us down this road.

Q: In the early 1980s, there were three concepts, as it were: prevention, adaptation, and mitigation. From the early eighties to the mid-eighties, you could not talk about adaptation because that meant surrender. All the talk was about prevention. Then all of a sudden, there was a policy shift at the Environmental Protection Agency or somewhere, and then the focus shifted to adaptation—we cannot do anything—and then to mitigation, and now adaptation is back in. I would like to see more talk about prevention. True, we can't prevent climate change from happening. It is changing already. But we can prevent people from doing things that are going to exacerbate it. We are talking about these coal-fired power plants. We are talking about building new things. There is a prevention aspect to climate change, otherwise, it is all adaptation—business as usual, and then try to stop it if you can and adjust where you can. Would you support the idea of bringing back the notion of prevention, or do you think it is embedded in these other words?

Morehouse: You are correct. I do tend to group prevention and mitigation together when talking about climate change. For example, capturing and sequestering carbon from coal power plants is mitigation, and also prevention. Given that human activities have already created a certain amount of climate change, I think of activities preventing additional change and mitigating its effects as the same. However, if this is a useful distinction, perhaps we should be more careful in our use of the terms.

Response from Questioner: I think this is part of the problem. A lot of people do not know that the IPCC and the UN define these terms in different ways. To me, when you mitigate things, you are proactive—you anticipate a problem and do things to prevent it. When you adapt, you are reactive. Something hits you, and you adjust. I am not saying that I am right and you are wrong, I am merely saying that we need to clarify our terms when we talk to the public to make sure that they understand what it is that we are saying when we talk to them.

Audience: I think when you are discussing adaptation and mitigation, mostly it has to do with "How do you make sure the effects are not as bad as you think in terms of energy and climate change?" It is leveling off those curves, as it were. But when we talk about how to have better governance, that, to me, is prevention. In other words, I think that to handle the effects when they occur is not simply adaptation and mitigation, but prevention of the effects on the people themselves.

Q: The Navy has been quite successful at using the seasonal forecast to adjust when you turn on and off heating and cooling. Is the Army using this?

Sheets: I am not a utilities expert—I come from the environmental side of the house—so I am not too comfortable answering this. I assume you are talking about main steam plants that provide heating and cooling to buildings where you have large installation size heating and cooling systems. Axiom has some, but I don't know what the Axiom policy is on how to handle these.

Audience: Anybody in the Army that has been a commander of an installation or a concern has faced the issue of turning the heat on too early or too bad. And, yes, it occurs, and, no, it is never done quite right.

Audience: The Marine Corp solution is "Too early, too bad, too late, too bad."

Sheets: I think they are trying to decentralize those units so the people who are in the rooms can do that—turn them off when they don't need them and then turn them on when they do. But I don't know that the infrastructure allows that right now in all locations.

Q: You talked about the economics of it. When you discuss these issues within your organizations, what kinds of costs do you look at? When you do total cost accounting, do you take into consideration what it costs when a gallon of oil is burnt in terms of, say, health problems?

Sheets: We do have health people looking at some of these kinds of effects. But I don't know that it is tied into sustainability quite yet. We have people working on all sorts of aspects of that. I don't know if the health effects of carbon dioxide have been included yet in the models.

Morehouse: We haven't included this in our discussions of relevant externalities that should affect the outcome of the requirements determination process. Occupational health issues come up in discussions about materials and processes used to manufacture and maintain systems because these result in direct costs to the Department and risks to the Department, given new international regulatory regimes such as the European Union's REACH legislation. But addressing the health effects of the carbon dioxide emitted by military systems caused by their use of fossil fuel is not something I think we could get support to consider at this point.

Q: Speaking of opportunities for military diplomacy, there is one that did not get mentioned. You mentioned AFRICOM. There are areas in Africa that have no power grids whatsoever. It might be useful to use some of these alternative energy sources to develop power grids. Nigeria, which is larger than the state of California, produces as much electricity or less electricity than the city of Washington, DC. We might be able to partner with NGOs to help them develop energy as a sign of good will.

I have a question for Karen Hulme. You did not mention things like environmental crime, environmental piracy, or environmental terrorism. Do you see a move in the international law community towards adjusting international law to look at nonstate actors/terrorist types of groups?

Hulme: I only covered the laws of armed conflict. I did not have the time to cover **intrastate** conflict. When it comes to noninternational law on conflict—and we are talking internal or civil strife, as we call it—there is no treaty. Well, there is a treaty—but there is no treaty **provision** when it comes to protecting the environment. You have to look to customary international law, which is pretty much what states do, what states say. Some codes in this body of law deal with what nonstate actors/militia are doing. As far as I know, none of the environmental provisions are binding in noninternational law on conflict. In noninternational law, the applicable law is usually simple criminal law. As to terrorism, we may speak of the Global War on Terror (GWOT), but "war" here is not war in a legal/technical sense. It is not armed conflict. And so the closest I think I can get to your question is to look at what Saddam did when he set fire to the 590 oil wells in Kuwait. This action was analyzed by many. The verdict was that it was not militarily necessary. It was destruction for the sake of it, punitive destruction. But from that, you saw the international reaction, and the international reaction was pretty much global condemnation. What we see is a kind of instant customary law in the making. I know that it was not a nonstate act. But it did make clear that the global community condemned the use of the environment as a weapon of war for reprisals. This goes back to ENMOD, I guess. But, no, when it comes to nonstate actors and the laws of war—I don't think it is there when it comes to the environment at the minute.

Schroeder: Can I just add a comment? As you know, our current definition of domestic terrorism is broad enough to cover an operation like Greenpeace setting up a blockade to prevent people from going whaling. It requires a threat or actual violence or damage to property with an intention to influence government policy. A provision of the Patriot act expanded it this way and got Greenpeace quite upset. But in the United States, the Justice Department has used some of the material support and other sanctioning provisions enacted in the Patriot Act to bring actions against domestic environmental activities.

Audience: I found this panel very instructive, and I take great heart from the activities that are being pursued and the opportunities that are available to achieve environmental benefits from reforming the peace time military. What I have a hard time getting my mind around is the possibility that the military services could ever be convinced to seriously alter their operational schemes in the name of environmental friendliness. That is, I envision these arguments between the war fighters and all of the rest of the "tail" who go around the world telling them that they should have fuel-efficient tanks or whatever. Do you think there is really any prospect, either on the logistical side or on the legal side, of actually altering their behavior when they are engaged in combat?

Morehouse: You are right. The convincing argument is not based on environmental friendliness. Efficiency is being seen as a way of delivering more operational capability from a more efficient force-structure. Since it reduces our logistics tail, it also reduces operational vulnerabilities caused by large soft fuel convoys that offer the enemy a target rich environment. The idea that we can deliver greater warfighting effectiveness with less logistical effort is appealing to war fighters. The selling point is not that we would reduce fuel consumption *per se* or reduce our carbon footprint. But it is something we could take credit for if we can measure it properly.

Sheets: Last night General Kern mentioned the Air Force Environmental Conference that went on a couple weeks ago. I was there. People there raised the point that addressing global warming would reduce dependence on foreign oil—and that is another big strategic issue for us. And I will tell you there are people who think they have that licked, and they typically come from West Virginia. They come from Pennsylvania. They think that they can reduce our dependence on foreign oil with processes like coal gasification and coal liquefaction—the Fisher Tropsch process. We haven't figured out how to get rid of that CO^2 from the coal. But this is something to think about.

Hulme: When it comes to changing the armed forces' behavior, as I mentioned, military pragmatism does play a part. A lot of the laws that are on the books are there because the military want them there. They want them because they want a disciplined, not a pillaging Army. The laws are not imposed on them from the outside. When it comes to operations, public pressure may be a big factor. War aims, too, may make a difference. If a conflict is about water or lack of water, for example, this may influence what the peacekeepers or what the military will do in the conflict. I mean, if it is about lack of water, they are not going into the armed conflict, hopefully, and just deplete the water even more, and make it even worse. So the aim will be important, too. And, of course, when it comes to the environment, the United States does not act on its own when it comes to military operations. It tends to look towards multilateral operations. So interoperability, working with partners in Europe and elsewhere when it comes to the environment, is an interesting issue that the United Kingdom (UK) and the United States want to protect more than others in wartime. One example I have—it is a lovely little example from the Balkan Conflict. The UK and the United States went to great effort not to breach an international treaty when they saved a brown bear to take it to safety. And so you cannot simply say that the environment has no priority in wartime. The UK and the United States are heavily influenced by it.

Q: I have one brief observation and one very quick question. The observation is that obviously the military is a large energy user, and any measures that can be taken to reduce that without compromising warfare seem to be tremendously valuable. But I think we should be careful about other claims about the military's impact. A lot of research done by the

International Union for the Conservation of Nature and others suggests that the environmental impacts of land managed by the military are considerably less than those of land managed by other actors. It further suggests that the environmental impacts of wartime are generally less than the environmental impacts of peacetime in most situations. So we should be careful about bundling too much together to demonize this.

My question is do you have any sense of what the energy use is of non-U.S. militaries and actors other than militaries, when they are engaged in similar types of operations around the world?

Morehouse: I don't know for sure. The energy intensity of military operations has increased over time, but I suspect the energy intensity of conventional military forces among wealthy countries would be comparable. But this would not hold true for developing countries or for asymmetrical threats. For example, an insurgent driving a Toyota to plant an improvised explosive device (IED) will have far more combat effect on a per gallon basis than a mine resistant ambush protected (MRAP) vehicle.

CHAPTER 7

THE U.S. MILITARY: PLANNING FOR THE FUTURE

Chapter 7 looks at what the U.S. Armed Forces might be called upon to do in the event that rapid-climate change does indeed take place. How should they plan for such a future? Professor Bert Tussing considers what the military can and should do in planning and responding to disasters and warns of some of the dangers that must be avoided. Major General William Nash brings personal expertise to bear on his discussion of peacekeeping and stabilization operations, and Dr. Douglas Johnson analyzes the strategic challenges facing our nation.

The Role of the Military in Civil Support

Bert B. Tussing
U.S. Army War College

The role of the U.S. military in disaster response operations, like the role of the military in many areas of "civil support," inspires conflicting responses in much of the civilian population. Civilians traditionally think of a U.S. general officer as someone with a triple-A type personality who arrives on the scene and takes charge. That is the sort of thing the American people like to see in times of crisis. At the same time, the American people grow weary, very quickly, of the military **remaining** on the scene, and particularly remaining on the scene "in charge." It is a part of the American psyche rooted in British experiences during the pre-Revolutionary War period, reinforced by the rise of the *posse comitatus* mindset following the American Civil War, and reaffirmed many times since.

The fact of the matter is, however, that no one is **more** aware of that psyche and its accompanying sensitivities than the leadership of the U.S. military—

from its flag and general officers to its civil masters in the Department of Defense (DoD). When we talk about "support to civil authorities," we mean precisely that; and nowhere in our doctrine or policy does "support" translate to "taking over."

Combat Operations.

The range of military operations shows how this mindset fits comfortably within our capabilities. At one end of the spectrum, the military is organized, trained, and equipped to conduct combat operations. In their ultimate manifestation, these operations are intended to fight and win our nation's wars, or leave the immutable impression that we are prepared to do so. Operations conducted to achieve these ends include the types of "kinetic" activities commonly and uniquely associated with the military: large scale combat operations, blockades, and so on.

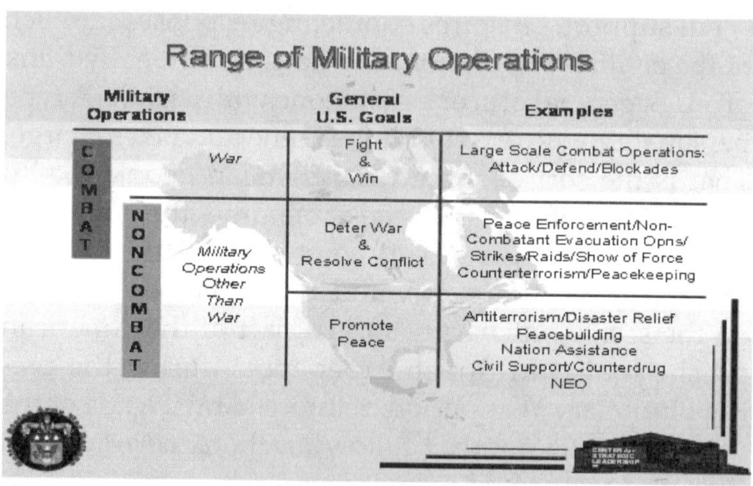

Figure 1. Range of Military Operations.

Military Operations Other Than War.

Noncombat operations, on the other hand, are not as cut and dried. These are manifested in what we have euphemistically come to term "Military Operations Other than War (MOOTW)." The "upper end" of MOOTW can be played out in support of deterring war and resolving conflicts. In these kinds of operations, our forces may move in and out of armed conflict, based on the necessity of mission requirements. It is important to note here that the military will always retain the ability to shift back to a "combat footing" to ensure force protection or to provide for the safety and security of a selected civil population. The complexities surrounding this requirement should be apparent. Further discussion, however, lies beyond the scope and intention of this section.

The "lower end" of MOOTW, devoted to promoting peace, deliberately avoids (but will never rule out) conflict. It calls upon a different kind of strength—the kind that the military typically applies in combat support or combat service support operations. The goal of these is to provide support to civilian authorities in a domestic or international scenario. Antiterrorism operations, noncombatant evacuation in a permissive environment, and civil support missions are subsets of these types of operations; but of particular concern to our intent are missions surrounding disaster response.

The resources and capabilities possessed by the military clearly recommend its use in disaster response operations, but not always in ways that are readily apparent to the casual observer. There are some obvious and logical ways in which military assets can be brought to bear in regions devastated by disaster and which have, temporarily at least, lost their normal ability to function. The military has significant engineering and communication capabilities as well as transportation assets. Similarly, the military has the ability to provide supplementary emergency health services until the civil component is restored. Less intuitive, however, are strengths that the military brings to the table that may be specifically tailored to aid in disaster response. The full breadth of the military intelligence community may not be applicable to needs of the moment, but the mechanisms within that discipline which allow for the collection, analysis and dissemination of vital information may recommend themselves. The fighting prowess of our special forces is well known; but less intuitive may be their value in terms of language skills and cultural awareness, unmatched in our "line forces." The competence housed in the military's civil affairs units is not something that immediately springs to mind, or something that the military is overly eager to offer for reasons that will be addressed later in this section. Nevertheless, the wherewithal contained in these units can fill a critical void in the absence of basic governance requirements and rule of law until the duly constituted authorities of a stricken area can be restored. And following catastrophic events which traditionally bring out the best and the worst in humanity, the element of security that the military brings to the field is simply essential—for without it, none of the other elements of restoration may be able to take root, or even be introduced.

Accordingly, one begins to see the benefits the military **can** bring to disaster response, but that does not begin to explain **why** the U.S. Government would want to see them applied to those ends. After all, at the end of the day there will always be some concern that committing the armed forces towards these altruistic ends may be a nice thing, but it diverts their attention away from their "day job" of protecting and defending the people of the United States, their values, and their interests. That defense (which frequently takes on a very offensive tone) can be done by no other organization; no reciprocal supplementation of capabilities can be effectively offered by any other segment of the government or the society it serves.

Such a simplistic approach, however, overlooks the ultimate strategic objective of defense, which is the security of our people, our values, and our interests. Without meaning to deny the genuine desire of the U.S. Government and its people to reach out to assist other parts of the world, our **National Security Strategy** reminds us that regional stability, in a world growing more and more connected, is inextricably tied to the strategic security interests of our own country. Accordingly, over time specific types of civil support operations have been developed, defined, and refined by the military. These serve as the armed forces' component of an interagency, and by extension international, means of promoting that stability. [1]

Civil Support Operations.

National Assistance Operations. One type of civil support operation is national assistance operations. This category is the broadest in scope of any we will consider. It includes military or civil assistance rendered by the United States to a nation during peacetime, crises, or emergencies (to include war), based on agreements mutually concluded between the nations. National assistance programs include, but are not limited to, security assistance, foreign internal defense, other DoD programs, and activities performed on a reimbursable basis by federal agencies or international organizations. It is included here inasmuch as it sets the stage for other programs more specific to our concern.

Foreign Humanitarian Assistance Operations. When we consider disaster relief, foreign humanitarian assistance operations are, perhaps, those operations which spring most quickly to mind. Conducted outside of the United States and its territories, these include programs conducted to relieve or reduce the harmful results of natural or manmade disasters or other endemic conditions. U.S. forces can draw on their appreciable capabilities to provide humanitarian assistance. However, this aid is deliberately limited in scope and duration. At first glance, this may seem arbitrary, or even callous. However, the purpose of these operations is to supplement or complement the efforts of the host nation's civil authorities or agencies that may have the primary responsibility for providing foreign humanitarian assistance. This reemphasizes a recurring theme in the military's role in these affairs: Regardless of the **scope** of its support, the military will **remain** in support, and not in charge under all but the most extreme circumstances.

Civil Administration Operations.

When those extreme circumstances do arrive, the chances are that civil administration units will be called in. This is a mission that has historically fallen to the military, but one that it does not necessarily welcome. The purpose of civil administration operations is to restore basic government functions, to include the provision of security and the rule of law. The very fact that the military is called upon to take up this kind of mission should mean that there are no other practical alternatives. Nevertheless, doing so casts soldiers in the ill-fitting role of policemen or statesmen, roles that inherently cast them in an overseer capacity that falls well outside of our notion of "civil support." Accordingly, the military enters into these commitments on the understanding that there will be an "exit strategy." It defines those steps and measures that will allow the passage of these authorities to a duly constituted civil structure. This structure may eventually take the shape of the stricken nation's leadership, or a recognized international body, but the intent of the military is clear: establish secure conditions for the passage of control, and then execute that passage.

The American military is also reluctant to take on extended civil administration operations because of the long-term effect such operations would have on the country receiving aid. Its approach here is similar to its approach to civil support operations in a **domestic** environment. Our uniformed leaders realize that theirs should be a "last in-first out" requirement. When the dust settles following a disaster, neither the United States nor (especially) its military, is interested in retaining control over the country they have committed to assist. Rather, we would prefer to restore to power appropriate civil authorities. They are the ones who, over the long term, will remain behind to provide for the governance and the well-being of their own people. The longer the military remains in control, the longer it will take these authorities to establish control and regain credibility in the eyes of the afflicted people. Remaining any longer than necessary, therefore, could hinder rather than help the appropriate restoration of authority.

Military Civic Action.

Military civic actions further reflect this attitude, albeit, perhaps, in a more subtle fashion. These missions, sponsored by U.S. forces, assist in projects benefiting local populations overseas. They are set up in such a way as to deliberately highlight the role played by the host nation's military. Operations of this sort may be directed at improving conditions in the fields of education, public works, agriculture, transportation, communications, health, sanitation, and others contributing to economic and social development. In these operations, the United States deliberately keeps a low profile, operating in the shadows as it were. Americans may advise the indigenous forces, and they may provide extensive logistical backing; but the most visible effort will be conducted by the host nation's military. The logic behind this approach has been borne out in Theater Security Cooperation Programs of every combatant command: properly planned, executed, and promulgated in close cooperation with local authorities and community leaders, military civic action projects will result in popular support for the military. As counterintuitive as it may seem, our intent in these operations is not to increase the popularity of the U.S. military among the local people, but rather to restore their faith in their own military. This can pay particular dividends in nations where the military has traditionally been seen more as an overseer of the people than a servant.

This is not to overplay the military card, however, not even the host nation's military card. If one were to ask any U.S. combatant commanders to recite the **Prime Directive** of civil support, he would likely declare that the primary responsibility for the well-being of a population rests with lawful government or the *de facto* authorities of the affected state. Domestic preparedness is first and foremost the responsibility of the civilian government and its agencies. This serves as the ultimate compass for the military in these matters, reminding them who is to serve as the directing force for the support that they bring, and who will remain to direct the extended recovery of a nation or region long after they have departed. At the same time, it alerts the civil authority to the beginning and end of the military's function.

Civil-Military Relations.

The relationship between the military, civil authorities, and the civil population they are both pledged to serve in disaster response has been carefully developed, scrutinized, and institutionalized in recent history. One could postulate that the modern origins of these relations evolved with the evolution of the concept of civil defense out of World War II. The immediate images inspired by those words may conjure up visions of "lights out drills" on U.S. coasts or even "duck and cover exercises" in grammar school during the early days of the Cold War. But the actual definition of civil defense operations was laid out in Protocol I of the Geneva Conventions. This referred to these operations as ". . . the performance of some or all of the international humanitarian tasks intended to protect the civilian population against the dangers, and to help recover from the immediate effects of hostilities or disasters . . . to provide the conditions necessary for its survival." The immediate intent of the protocol, of course, was to protect the most vulnerable from residual effects of battle, but parallels were easily drawn to combined civil-military operations in support of populations suffering from the effects of natural disasters.

One of the difficulties faced in these combined operations, of course, was in closing the gap of cultural differences between the way the military approaches a problem and the way their civil counterpart, partners, coordinators, or even directors addressed it. This has led to what can be described as an inherent complexity in these procedures. The innate efficiency the military most often demonstrates in the accomplishment of its mission is built upon what it refers to as "unity of command." This unity is foundational in the armed forces; it immediately removes the preponderance of questions over who has authority to direct actions, control resources, and prioritize efforts. Its purpose is described in our doctrine as a means to ". . . ensure unity of effort under one responsible commander for every objective. Unity of command means that all forces operate under a single commander with the requisite authority to direct all forces employed in pursuit of a common purpose."[2]

The problem with unity of command is that it can only be achieved in what we might think of as a recognized chain of command. These concepts become almost immediately unhinged when you introduce elements into response and recovery operations that are not in the chain, and are not inclined to insert themselves in the chain. At that point the goal shifts from trying to achieve unity of command more directly to its purpose—which is ensuring unity of effort. One can quickly discern that, without the immediate authority embodied in our concept of unity of command, this becomes a far more complicated effort. The complexity is even forecast in our doctrinal definition:

> Unity of effort, however, requires coordination and cooperation among all forces toward a commonly recognized objective, although they are not necessarily part of the same command structure. In multinational and interagency operations, unity of command may not be possible, but the requirement for unity of effort becomes paramount.[3]

With no intent to sound parochial, the more diverse elements are introduced in disaster response operations, the more difficult achieving unity of effort will be. In the U.S. military's *Joint Doctrine for Civil Military Operations* (Joint Publication 3-57), the military commander is specifically reminded that "[a]chieving unity of effort becomes more complicated with the increasing number of nonmilitary players (e.g., NGO's [nongovernmental organizations], international organizations, other government agencies, and UN [United Nations] agencies) involved in operations." This in no way implies that the military is seeking means or approval to "go it alone." It only points to an inevitable challenge that will have to be overcome if the combined civil-military relief effort is to effectively (if not always most efficiently) provide for the victims of a given disaster.

Again, it may be counterintuitive to those who expect every military response to reflect the subtle nature of a Patton, a Sherman, or a Genghis Khan, but DoD approaches the challenges of blending civil-military efforts and capabilities as very much of a give-and-take affair. Illustrative in this regard is the Department's approach to working with NGOs. DoD is keenly aware of the tremendous capabilities NGOs bring to disaster relief operations. Frequently associated with the areas impacted by the event prior to the event, the NGOs have access to information and background that no governmental organization, certainly to include the military, may bring to the fore. However, the very nature of NGO interaction with the population of these areas may not recommend their close association with any individual government (particularly the United States) due to cultural sensitivities and political nuances. Candidly speaking, such associations could close many doors to the NGOs that would otherwise remain open. These sensitivities become even more pronounced when those associations become aligned with military organizations, particularly in areas of the world where their history involving militaries has more often taken an oppressive tone.

Aware of these backgrounds, the U.S. military recognizes a need to take a more balanced role in dealing with NGOs, with the "balance" leaning toward less visibility. As pointed out in the aforementioned *Joint Doctrine for Civil-Military Operations,* the military "needs to understand and try to facilitate the principles of NGO's operational and financial accountability." Moreover, our uniformed leadership needs to recognize that the appreciable capabilities they bring to relief operations are not the only capabilities available for those ends, and frequently not the most affordable. Civil agencies (governmental and nongovernmental) will often possess comparative operational advantages for emergency relief work. They will frequently be able to do so at 10 times less the cost than it would take for the military to perform the same or similar functions. The Army is very good at providing all of the capabilities that have been referred to throughout this discussion. However, the total package means not only providing manpower and resources. It also means sustaining these capacities and retaining the kind of cohesion and integrity that will allow units to transition (as necessary) from noncombatant operations to combatant operations. All of this can come with a remarkable price tag.

And so the military attempts to remain aware of the constraints, limitations, and occasional allowances that are necessary in working with its civilian counterparts. But there is an accompanying set of sensitivities those nonuniformed counterparts must also keep in mind in working with the military. To begin, the military—always aware of the potential requirement of having to return to their "day job"—has certain tests of legitimacy that will be weighed against any request for support. Generally speaking, these tests are:

- The **legality** of the request.
- Issues of **lethality** which become far more pronounced the further away the uniformed element moves from the battlefield into the civilian population.
- Accompanying **risk** to the soldiers performing the mission and the victims they are serving.
- The **cost** of the required resource and manpower expenditure.
- The cost of the expenditure in terms of **readiness** to resume the missions only the military can perform.
- Whether or not the task is **appropriate**—this subtle point can only be determined by the commander who is in charge of the personnel and resources.

These tests for legitimacy for military inclusion should become a part of the civilian planning process. In that process, as much as is feasible, the viewpoint of the military operational commander whose units' capabilities are being sought should be solicited.

It is not hard to imagine in these complex environments that civil military operations may not always proceed smoothly, in spite of cooperative efforts on both parts or, especially, in the absence of the same. As if to illustrate this point, some international organizations and NGOs have demonstrated a traditional reticence about sharing information with the military that could prove vital to disaster response and recovery operations. Nevertheless, our doctrine reminds the military commander that it is most often his or her responsibility to "go the second mile," and do as much as possible to cultivate a climate of respect and trust. This will facilitate vital exchanges between our agencies, not for the sake of the agencies, but to help an afflicted population. Sharply differing civilian and military perspectives usually work against achieving a common operational culture. But, returning again to our recurring theme, military acceptance that civilian tasks in many modern emergencies constitute the main operational effort—and that military tasks are in support of this main effort—can assist greatly in achieving a compatible culture in the field.

Contributions of the U.S. Military.

In spite of these difficulties, the U.S. military, in collaboration and coordination with civilian authorities, has made—and will continue to make—significant contributions in the area of disaster response. These notably include initiatives designed to mitigate or preclude natural disasters. The U.S. military, in fact, is taking an active role in environmental security issues around the world. These efforts are orchestrated through the office of the Deputy Undersecretary of Defense for Installations and Environment and are taking place in nearly every combatant command. Consider, for example, the U.S. Southern Command's (USSOUTHCOM) training initiatives surrounding environmental security in Paraguay; efforts in the U.S. Pacific Command (USPACOM) to predict and mitigate the devastating effects of seismic disasters in Southeast Asia; and the U.S. Central Command's (USCENTCOM) initiatives to address the looming threat of water shortages in many areas of its theater of operations.

Obviously, these are not initiatives that DoD is taking on alone. DoD's chief interagency partner in these endeavors is the Office of Foreign Disaster Assistance (OFDA) of the U.S. Agency for International Development (USAID). In addition, the combatant commanders have the means of directly coordinating their efforts in these regimes with the U.S. State Department's 12 Regional Environmental Hubs, whose focus is aimed at promoting regional environmental cooperation, sharing of environmental data, and adoption of environmentally sound policies that will benefit all countries in a given area.

Beyond these "senior partners" in the endeavor, DoD frequently collaborates with other members of the federal interagency process focused on these issues. Included in their ranks are the National Oceanic and Atmospheric Administration (NOAA), the U.S. Geological Survey (USGS), and the Environmental Protection Agency (EPA). Outside of our own governmental agencies, DoD's efforts have frequently been coordinated through organizations like the United Nations Office for the Coordination of Humanitarian Affairs (UNOCHA), the International Search and Rescue Advisory Group, and (of course) the Ministries of Defense and Environment for countries that have been victimized by disasters of the scope and severity envisioned in these discussions.

The Logic behind the Efforts.

The impetus behind these efforts on the part of the U.S. Government in general, and its military in particular, is not totally altruistic. More and more we are awakening to the realization that regional stability is directly impacted by environmental issues. Droughts in Central Africa signify dangers and suffering that will not be contained by political boundaries. When catastrophic events, like the earthquake in Pakistan, the tsunami off of Indonesia, or the hurricane in Honduras, wreak havoc on congested areas, they can destabilize an entire region. Ecological disasters like that we see in the Caspian Sea region threaten to trigger a cataclysmic series of events that could result in human suffering, deprivation and, ultimately, conflict. Doing something now can head off trouble later. Better community planning may prevent migration crises. Exposing, understanding, and preventing the neglect or deliberate abuse of the environment can fend off disaster. In short, environmental disaster is being recognized as a precursor to regional instability; and in today's world of globalization, regional instability is a precursor to global instability.

Fortunately, realization of shared threats and shared concerns has led to initiatives towards regional cooperation in every combatant theater. These initiatives focus on shared means of identification, prevention, mitigation, and (when required) response to both manmade and natural disasters. For almost a decade, the U.S. Army War College's Center for Strategic Leadership, under the direction of Dr. Kent Hughes Butts, has partnered with the combatant commands, USAID, and organizations like the Woodrow Wilson Center's Environmental Change and Security Program, to bring together regional representatives around the world to address these issues.[4]

The types of lessons that have come out of these and similar efforts have reinforced the role of the military not only in disaster response, but in disaster preparedness. Theater Security Cooperation Programs (TSCPs) sponsored by all of the combatant commands have planted the notion that in addition to their traditional skill sets in communication, logistics, and security, the benefits derived from the military's proficiency in planning and strategic policy development further recommends them to these efforts. A vital component of the combatant commander's resources towards these ends is his Joint Interagency Coordination Group (JIACG) which houses governmental and, occasionally, nongovernmental representatives from outside of DoD. These representatives serve as key advisers to the combatant commander in understanding and coordinating the combined capabilities of their organization, identifying synergies and efficiencies, and ensuring to a far greater degree their optimal employment in times of crises.

In the past, the military has looked upon disaster response as a necessary function, but not one that was particularly welcomed. The concern over diverting attention, manpower, and resources away from the "traditional role" of the uniformed services no doubt served as the foundation for that reticence. But times have clearly changed. Environmental security and disaster prevention, response, and recovery are now looked upon as acceptable military missions in that they are viewed as essential elements of regional stability. Specific programs devoted to these concerns, such as USPACOM's Disaster Preparedness Mitigation Assessment program, USCENTCOM's Disaster Preparedness Program, and USPACOM's Disaster Relief Response Process are taking shape in every theater.

International Efforts.

In emulation of the American effort, other countries' governments and their militaries are seeking their own civil-military synergies. Most governmental organizations responsible for environment and disaster response are relatively understaffed and under resourced, lacking manpower to undertake the activities necessary to efficiently accomplish these broad-based missions. The military, by working with these organizations, allows them to accomplish a great deal more. In the process, militaries that have been a source of suspicion and fear in earlier times are being rightfully cast as servants and protectors in the eyes of their own people.

Finally, developing regional response capabilities to address a set of common concerns may serve as a foundation for other cooperative endeavors. Countries that would otherwise never gather around a negotiation table are currently engaged in regional efforts to forecast, mitigate, and if necessary respond to cataclysmic natural disasters. Constrained resources have themselves provided the impetus for greater cooperation, and a judicious division of labor in preparing for these calamities is being directed at producing synergies and eliminating redundancies in providing for the common needs of a region's citizenry.

As is frequently said, much has been done, and much remains to be done. But as the U.S. Government attempts to come to terms with an evolving list of environmental issues, its military must be counted upon as an essential component to any solution. Before an event, the military represents the most tested entity for operational and strategic planning in the country's inventory. During the event, our uniformed forces can assume a responsive posture that cannot be duplicated in terms of manpower, capacity, timeliness, or capabilities. Following an event, the armed services will be hard to match in terms of what they bring to the field in restoring essential services, and setting the conditions for long-term recovery. But little of the military's vast potential in meeting these demands will be effectively brought to bear without understanding on both sides of the civil-military framework of how it should and should not be employed. In the final analysis, the responsibility for that decision lies solely in the camp of the civil authorities.

ENDNOTES - Tussing

1. They include engineer, transportation, communications, health services, intelligence, special forces, civil affairs, and security.

2. Joint Forces Staff College Publication 1.

3. *Ibid.*

4. Some representative examples are U.S. Southern Command, Regional Environmental Security Conference, Asunción, Paraguay, May 2002; U.S. Central Command, Consequence Management Conference, Washington, DC, September 2003; U.S. Central Command, Disaster Response Conference, Almaty, Kazakhstan, October 2003; U.S. Central Command, Disaster Response Preparedness Conference, Manama, Bahrain, September 2004; U.S. Southern Command, Environmental Security Training Workshop, Alajuela, Costa Rica, January 2005; U.S. PACOM and U.S. CENTCOM, Seismic Disaster Preparedness Conference, Honolulu, Hawaii, February 2005.

Peace Operations

Major General William L. Nash, USA-Ret.
Council on Foreign Relations

I really did not think this climate change issue was especially important until I got my *Sports Illustrated* a couple weeks ago. And I found out that between 1954 and today the mean temperature of the United States has increased to an alarming degree. Indeed, had the temperature then been what it is now, the fly ball that Vic Wertz hit in the 1954 World Series would have gone an additional three or four inches due to the temperature of the air, and Willie Mays might not have made that miraculous over the shoulder catch. Worse yet, the same temperature changes facilitated the movement of the Emerald Ash Borer Beetle to Pennsylvania and New York, where that little devil is boring into the ash trees from whence the best wood for baseball bats come. This stuff is serious, folks. Maple shatters too much, and they are switching from ash to maple and that is not as good. No, sir. The only thing worse than maple bats are aluminum bats.

The problem with being the next to the last speaker, before the sum up, is that almost everybody has already said most of what you planned to say. As a result, most of the points I make in my presentation will reinforce ideas you have already heard rather than say something entirely new. But I submit to you that I *am* one of the few guys in the room who has actually conducted or participated in a military operation for the specific purpose of improving the environment.

In August 2000, North Atlantic Treaty Organization (NATO) forces in Kosovo (KFOR) and the United Nations (UN) mission in Kosovo conducted the takeover of the lead smelter in Zvecan, Kosovo, in Northern Kosovo. I was a civilian administrator in charge of the civilian side of this operation. We took over this plant because of the pollution that it was putting into the air throughout the area. We also hoped to gain economic opportunities through refurbishing this disastrous site. And I will tell you how bad it was. We took a blood test sample of 100 people, including myself and 99 locals. I had been in Kosovo for less than 6 months. I was the only one whose lead content was within the normal range—and I was at the upper limit. Of the sample, 25 percent were 4 to 10 times the safe range for lead content in blood. This was clearly a nonethnic issue in that the Serbs and the Albanians were equally affected. The Romas, however, because their refugee camp was immediately downwind from the plant, suffered not only from the air quality, but from the lead soot that came down on top of their camp.

So we took over this plant in a military operation which was immediately followed by civilian authorities that had the expertise to shut down the plant safely. It took us about 6 hours to cool down everything, to establish an unemployment compensation system for the workers that were being put out of work, and to facilitate the departure of a number of thugs that ran the place; they went scampering back to Belgrade very quickly. But we took over the place. We then drew up a contract for $15 million to bring in a Swedish mining firm to conduct an environmental cleanup and a feasibility analysis to restart production. The economic potential here was that the mines were largely in Albanian areas. The smelters and the processing facilities were in the Serb areas of Kosovo. They were looking for an opportunity for mutual benefit to put them together.

Two days after the takeover, I went to a meeting in the Serb Union with 100 guys. It was a little tense. My security detail took their weapons out from under their jackets. They parked my car outside the window so that I could jump through if things got really tense. And the Serb leader, thinking that we had taken over the plant to give it to the Albanians, stood up and said, "Are you going to let those Albanians come into this plant?" And I looked him in the eye and said, "You're darned right we are. They are going to be driving the trucks bringing in the ore. And you guys that run the smelter can do your jobs." And they said, "Well, that's fine. That's fine. We would like work. We would like prosperity, we would like an opportunity."

Of course, the end of the story is that we couldn't pull it off. That was because the economic analysis didn't come up with sufficient profit margin to make it happen. Of course, they didn't count the unemployment compensation we were paying. They didn't count the added security cost, multiplied by years, because of the failure to do the refurbishment. They didn't take into account the political foundations that joint multiethnic economic associations would bring.

This is going to be the world we face in the 21st century and beyond in an era of climate change. We are going to find ourselves on many occasions in circumstances in which peacekeeping, the environment, politics, and social issues, all combine in a new formula for security.

I think it is very important that we talk about peacekeeping—or stability operations, if you prefer. The fact is that dealing with the consequences of climate change is going to be hard enough for the United States, for Britain, and for the European Union. For fragile states, it is going to be almost impossible.

So obviously we must anticipate the possibility that fragile states will fail, or that regional tensions will be exacerbated to the point of conflict. Given that these possibilities abound, I think you can see that the armed forces of the United States and her allies will have an increased opportunity to conduct stability or peace operations in the coming years.

As we look at this situation, I would like to throw in the fact that stability and peace operations should be viewed as a continuum. And I want to put my two cents in about the prevention, the mitigation, and the recovery or post-conflict reconstruction. You will find that the skill sets that deal with one aspect also apply to the other two. So, as we look to organize and train ourselves to conduct these operations, we will find that we can use the same skills, assets, resources, and wisdom before, during, and after conflict. I can make the argument that it is more efficient to do it in the prevention phase, as opposed to the mitigation or the recovery phase. But the whole capability is necessary. Today we are talking mostly about the recovery or the post-conflict environment.

Given all this, we should expect that we will be called upon to do both more and better work in this area. On the quantity side, whether by invitation, intervention, or invasion, I think you will see us having to engage in peacekeeping with greater frequency. I do not necessarily mean that we will get involved in another Iraq or another Afghanistan or even another Kosovo or Bosnia. We might well engage in much smaller, less visible efforts, dealing with a variety of problems. In addition to traditional peacekeeping operations, we might, for example, be asked to provide security assistance, to assist other militaries and other governments as they themselves struggle to deal with the consequences of climate change.

Requirements might increase, and this, by the way, is what Bert Tussing brought up about the Special Forces' capacity to teach and train. If migration becomes an issue, we will need to improve our capabilities in certain areas. We will need to be able to conduct surveillance of long borders. We will need to be able to understand what is happening and be prepared to deal with people crossing the borders so that we can establish refugee camps, at a minimum. We need to be able to survey large areas to find out where displaced people are so we can assess their needs, and then push forward and get them the supplies they need. How will we do these things? And how will we integrate our efforts with those of the local military and their government?

We are going to be concerned with lessening our own impact on the environment. I do not know if we can achieve carbon neutrality on these types of interventions. But we can go many steps in the right direction. We were talking earlier today about what motivates the military. We talked about how the military tended to overlook environmental concerns if it meant it could fight better. Well, let me tell you, as an armor officer, the worst thing that I could do is either run out of gas or let my batteries go dead. So, fuel efficiency is a combat multiplier for me. While getting ready to go to war in Operation DESERT STORM, I did not worry about anything as much as I worried about the availability of fuel. And I spent more time than I wish to remember writing memos to my bosses, drilling my commanders, and drilling my soldiers on how to refuel quickly so we could turn around the trucks and get them filled and get them to catch up with us.

Fuel efficiency is crucial. In the absence of an alternative, the only way to keep the batteries charged is to run your engines. Indeed, the only way to stay warm is to run your engines. Now, the M1 tank has got an auxiliary power unit (APU). But it is about the only vehicle other than an aircraft that has got it. Okay, Strikers are getting one. Why not the Humvee? Take, for example, a peacekeeping operation. Such operations are typically check point and observation point heavy. Think what this involves. Four Humvees stationed at a road junction for 24 hours a day. Check points at 100, 200, even 500 road junctions. Four Humvees running, keeping warm, keeping the radios working, at every one of the 500 road junctions—that is a lot of fuel. There is a way to reduce that. Incidentally, there is a way **you** can demonstrate **your** interest. You guys can go to Capitol Hill and get the police to stop running their cars all day right around our Capitol (and my neighborhood). In any case, it is an issue that is very, very important.

But the majority of the tasks that will fall upon us in disaster situations or peace operations are civilian in nature. They will include things ranging from governance to economic development, to social reconciliation, and to the development of the rule of law, which is beyond public security. It is the police, the courts, the justices, and the jails that will be needed.

Now, we will have to expand upon the advances that we made in the last couple years. The news about National Security Policy Directive 44 (NSPD44) and DoD Directive 3005 have brought some improvements. At least our government has laid down the principle that we need to achieve better civil-military integration—and I mean integration, not just cooperation; the latter is not good enough.

As Robin Dorff said earlier, the National Security Act needs to be revised. The first time I gave this speech, it was to say we need a new National Security Act of 1998. Now I am up to 20XX. This is a serious point, though. Look at 1947. That led to a dramatic change. The National Security Act of 1947 created the U.S. Air Force. It created the Central Intelligence Agency. It created the framework in which we have worked for 60 years, until the last couple of years when we have started changing the structure of our government. But we are changing it in very piecemeal fashion: a little tinkering with intelligence here, a little creation of homeland defense and a little transformational diplomacy there; that type of thing. We need to go well beyond what we have done today. And this begins with a larger definition of security in general and national security in particular. This definition must be much more holistic, much more comprehensive. I have not talked about taking all these issues and putting them to work with international and regional organizations. That is an additional requirement that we must think through as we go forward.

In the State of the Union address this year, the President called for the expansion of the Army and the Marine Corp by 92,000 people. In the same paragraph, he called for the expansion or the establishment of a civilian response corps. The budget, however, did not include any requests for funding such an organization. But, I raise that with you so you can think through how we make the State Department's coordinator a real entity. Creating a new director of foreign assistance, an administrator, is not an all-bad idea of how to integrate the State Department with the USAID. But how do we make a reality out this position? How can we develop an Inner Agency team capable of working along side the military in the stability and peace operations? All hard questions! Thank you very much.

The Strategic Challenges of the U.S. Army In the Face of Global Climate Change

Douglas V. Johnson II
Strategic Studies Institute

Introduction.

My talk will focus on the U.S. Army and the strategic challenges posed by global climate change. I want to begin by dropping into the structural mode for a couple of minutes to make sure everybody understands the landscape. There are a couple of key documents of which you should be aware. First, there is U.S. Code, Title 10. This tells the military services what they are directed to do and directed not to do—in other words, what the limits of their activities are. On top of that, or parallel to it, on an annual basis, the President of the United States is supposed to issue a National Security Strategy document. In this document, the President, his entourage, and the National Security Council survey the world, look for threats to National Security, and prioritize them. They do not necessarily go far beyond that. These are public documents and you can read them online anytime. If you do this—and if you teach National Security issues, it is an exercise you ought to put your students through—start with the early National Security Strategy documents and progress through them. You will see dramatic changes in style and variations in intent. Some are very self-congratulatory, "Look at what my administration has done." Some actually do what the law requires.

Thanks to our last Secretary of Defense, Donald Rumsfeld, we now have a new document called the *National Defense Strategy*. It is supposed to translate the National Security Strategy into military terms. It further prioritizes things, but it also begins the process of apportioning resources in accordance with those priorities and assigning general tasks. In the past, the Chairman of the Joint Chief of Staff has issued a National Military Strategy document. Whether that will continue remains to be seen. There is some overlap between that and the Secretary of Defense's document.

Those documents tell the military commanders in the field and the Chiefs of the Services what they are supposed to do. Notice that I just mentioned two different categories of people: the Chiefs of the Services and the commanders in the field. The combatant commanders in the field are responsible for what goes on in a geographic region, for the most part. The Service Chiefs are expected to raise, to train, and to equip the forces that are handed off to the regional commanders to perform the actions described in such documents as the *National Security Strategy, National Defense Strategy,* and *National Military Strategy.*

The Environment and the Army.

So, the military services are permitted and directed to act only within a very disciplined process. That there is an Army Environmental Protection Institute (AEPI) that exists within the legal structure of the service is indicative of authorized interest. It shows that environmental issues are being taken seriously. The Army does not spend money on things in which it has no interest. It has lots of things that it needs money to support. Investing funds in the AEPI is, therefore, signal number one that we are actually interested in this business.

Last year another document was published: The Department of Defense (DoD) Directive 3000.05. The number may not mean much to you, but it tells the armed services that stability and support operations, which include issues of environmental impact, will receive priority equivalent to combat operations. The Directive does not say much beyond this, so much of it is open to interpretation. Nor does it say that the same amount of money must be spent on these kinds of operations. And expenditures in the environmental category are, unless I am much mistaken, infinitesimal—$500 billion are to be spent primarily on combat operations. Nevertheless, this directive does send a message that the U.S. Army is interested in things happening on the environmental front, outside of the combat theater.

There is another way to look at this, moreover. Combat operations receive a great deal of attention and they cost a lot, but it is actually combat operations support that costs most. It is here that we can achieve most positive results. By working to reduce the harmful environmental effects of operations, we can also plausibly enhance combat efficiency.

On the battlefield, there are certain realities that are very difficult to deal with. For example, the Army was told that lead pollutes. So we took lead out of bullets. How much pollution do bullets really inflict on the environment outside of a rifle range? But we did it. We do not have lead bullets anymore. Well, yes, we do have them and there are places where they are needed, but we did back off from their use. This was done to avoid pollution. But on the battlefield, you want to be able to knock a guy down when you hit him with a bullet so that he doesn't keep on running and get close to you. For that, lead bullets are best.

The Army has not quite figured out what to do with this new DoD directive. It has one thing, though, in its kit bag which will enable us to do a lot of good things. This is a structure called a Maneuver Support Brigade. This is a headquarters which can bring under its control any grouping of entities to perform almost any essentially noncombat task that is assigned to that particular commander.

So, in one sense, the Army is a bit like a lady-in-waiting. We have a structure, of sorts. We have a directive of sorts. But at the moment, climate change is not placed high in the national scheme of priorities. So we cannot dedicate large amounts of money to this mission nor develop suitable structures. When Iraq eventually cools down—and we should probably also assume that it will take the Army some time to recover—I would predict, and I would hope, as part of the long-range vision, the potential impact of climate change will achieve some substance beyond a piece of paper and a diagram on a piece of paper, which is about where we are right now.

Potential Army Roles and Missions.

For us, disaster relief is a reaction mission. Disaster happens or we see disaster about to happen, and we go out and take care of the problem. Disaster relief and national security are not necessarily, or indeed often, synonymous. National security should be thought of as a proactive long-range undertaking within which disaster relief is one of those lesser-included things. It is vital that you separate national security and disaster relief in your thinking. When Mt. Pinatubo erupted, how much warning did we have for that—a week at most?

Now that we have volcanologists giving us information, we have years of warning. So new possibilities have arisen. And in this area of long-term planning, there are indeed ways in which the military can play a useful role in environmental protection.

The armed services are constantly scanning the world from space, from high altitude aircraft, and in various other ways. The first wave of satellites were primarily used to collect military data. Now, however, there are more satellites up there producing earth data than producing military data. And there are people who understand that feedings from these satellites—especially multi-spectral or geo-focused ones—are capable of producing excellent information which could warn us of disasters, both imminent and creeping. The military can participate in this. It is not necessary that all that information be classified and sequestered behind the "Green Door" (a colloquial phrase used to describe the pervasive barrier between useful intelligence and the people who need to use it.) Probably 80 percent of this information is shareable.

We also do planning very well. We train people to be planners. The staffs of each one of the regional combatant commanders have to deal with the prospects of environmental issues in their natural course of events. Can that be done at a national level as well? Absolutely. Suppose we were to write a National Security Act of 2010. I see no reason why we could not use it to heighten awareness of those kinds of environmental issues which pose a threat to the integrity of the United States first, and then its allies and friends, in that order. If we are indeed facing—as everyone now thinks we are—inundation of the coast and low-lying waters; if 20 percent of Florida is going to go under water as the sea level rises, should we not be seriously interested in this? Do we actually believe that this will happen? Can we begin to plan for this? It does not matter the color of the suit that does the planning.

One would hope that in this National Security Act of 2010 (which, by the way, has been on our plate for about 15 years) we might mandate, as an act or law, a means of integration of climatological information across government agencies. We created the Central Intelligence Agency (CIA). Cannot we write a law that says this intelligence agency can talk to environmental people? We created the National Security Agency. Can we not give it new properties which make it something other than the personal pet of the present president? Do we want to establish some connecting mechanism that will allow us to talk to nongovernmental organizations, private voluntary organizations, and people like that who are, as mentioned, not really comfortable getting close to military guys? Why not?

There is an awful lot of interest in doing this. What if environmental security were to become a part of the National Security agenda and become incorporated into a National Security Act of 2010? Can it happen? Absolutely! Can we get it to do the right thing? Absolutely. How do we go about it? The guys in uniform understand the need for this. The average guy for whom you vote, either in local elections or national elections, hasn't got a clue. So, what I am telling you is that there is a possibility of bringing all of this together in a synergistic fashion in about 6 years. Maybe. Is that too late? I do not know. Will the threats be so palpable then that they can be incorporated in this National Security document? Arguably, yes. How? To whom do you speak on a day-to-day basis? With whom do you correspond? Do you use emotion, or do you use facts when you address these issues? Can you establish clear linkages between actions and consequences? When you can do that, then we have a path to gaining control of this thing. Thank you.

Commentator

Richard J. Kilroy, Jr.
Virginia Military Institute

It is a pleasure for me to be here and be part of this discussion. By bringing together the military and academic communities, this conference is doing something that is really very important. The academic community tends to think of things theoretically. The military thinks of things practically. And sometimes the two communities do not communicate to each other very well. So this type of conference allows us to get to the heart of some of these issues. I must say, too, that I very much appreciate the opportunity to be on a panel with these three gentlemen, one of whom—General Nash—I have known since 1998. They are all practitioners, but they are also academics.

We are, as this panel stresses, being challenged with a new security environment. But in many cases, the military missions themselves really have not changed. We are going to continue to do things that we have historically done in the past: peacekeeping, humanitarian intervention and civic assistance, homeland defense, and homeland security. What has changed, rather, is the level of intensity. A couple of years ago at East Carolina University, a guest speaker addressing our cadets at an ROTC commissioning ceremony, stressed that the future would be a very different one. He made it sound a bit like Mr. Toad's Wild Ride at Disney World.

One question that has arisen over the course of the conference is whether we should focus more on threats or capabilities. I was involved with the U.S. Training and Doctrine Command (TRADOC) after the

Berlin Wall came down. We were trying to develop threat doctrine for training purposes; we went back to something called dial-a-threat. At the time, there were no more bad guys: You couldn't fight against the Russians, so they became the Crasnovians. We started creating the Marcollens. We created, basically, capability-based scenarios. We sought to put the military into challenging situations so they would have to exercise traditional roles and missions. But we actually could not identify what the threats would be. We are in a situation today that is not dissimilar. As this panel made clear, from a practitioner's standpoint, even if the military is not entirely clear what the threat is, it must still be prepared, trained, and equipped to perform its missions.

Let me turn to Professor Tussing's talk. He made some excellent points about capabilities. He is right that a lot of people do not think about what the military brings to the fight. But when they find themselves in one of these difficult environments, the military suddenly takes on increased significance. Even when the military is overseas and plays a subordinate role, a lot of civilian agencies are going to turn to the military, because it is the military that has the capabilities. He is also right that when the military works in foreign operational environments, it has to give credit to the host-nation militaries, to the host-nation governments. I served in Latin America for a number of years. General Clark used to say, "In SOUTHCOM, take no credit and expect none." And I think that was a good rule of thumb.

Professor Tussing also provided us with a lot of good examples of what combatant commands are doing today, whether it be in SOUTHCOM, PACOM, or CENTCOM. They are all looking at aspects of disaster preparedness and disaster response. How can we prepare to operate in those environments? We have to deal with the effects of climate change. But that is difficult when you have many of these high-demand, low-density items and assets, and especially when you are in a theater where you may not have the right kind of forces routinely assigned to you.

In his presentation, General Nash spoke of his own experiences in Kosovo. He applied some of his information operations (IO) training and stressed the importance of shaping perceptions. And, he said, if we can create win-win situations, especially when it comes to environmental issues, then we can accomplish our objectives. And I thought the example that he gave showed just that. He also made clear that peacekeeping in an age of environmental change has to take into consideration many different factors—the social, political, and cultural. And we have to integrate the different challenges, especially in areas like the Balkans.

It seems to me that the more we do on the preventive side, the more tools and resources we will have to respond effectively when the need arises. And General Nash's experience bears this out. When I taught IO, one of my major challenges was trying to convince warfighters that if we did IO well, they would not have to go to war. They didn't necessarily like to hear this. Tankers, you know, want to go out and kill things, right? But the fact is that doing IO well means that we do not have to put U.S. forces in harm's way and risk blood and treasure. So, the key here, again, is that the United States has to be able to master its own capabilities.

I also agree with our panelists: The U.S. military needs to become more energy efficient. We have to set a good example and do what we expect host nations to do. When I was in SOUTHCOM, it was hard for us to preach about counterdrug missions to host-nation militaries because we didn't do the same at home. We were speaking, as it were, with forked tongues. As General Nash says, the civilian leadership will play the lead role. But the best results will be achieved if an integrated approach is used.

Dr. Johnson also made a number of very important points. By going into the structural mode, he reminded us of how the military actually works and under what constraints it operates. This is a key point which we should bear in mind. Combatant commanders have the warfighting role and responsibility, but so do the Service Chiefs. Combatant commanders may be sold on environmental security and want to perform the kinds of missions we discussed. But they also have to have the ability to organize, train, and equip the necessary forces. And if the service components aren't read into it, it is very difficult to get that message out. On a related note, Professor Johnson also stressed the need for continuity between National Security Strategy, Defense Strategy, National Military Strategy, and so on. Here again, if the Service Chiefs are not involved, combatant commanders are not going to be given the resources needed to perform environmental missions. And in the end, it really comes down to whether or not the mission is funded.

Dr. Johnson also talked about planning. Combatant commanders plan. There are official documents which tell them what to plan for. The Unified command plan dictates areas of responsibility, missions in areas that the combatant commanders have to be prepared to do. SOUTHCOM had counterdrug as its warfighting mission. But it also had a heavy component of humanitarian civic assistance. We had to prepare functional plans to do those types of missions. And I think that Dr. Johnson is absolutely correct when he says that we need to ratchet up this kind of activity.

Before I close, I would like to make some general observations about environmental threats. When I was in SOUTHCOM in 1996 and 1997, I served as a special assistant to both General McCaffrey and General Clark. General Clark was trying to be very proactive at the time in environmental security. He tried to bring this within the scope of the combatant command. It was hard for him to get the attention of Washington. They said, "Counterdrug is what you do in SOUTHCOM, do not talk to me about disaster relief or environmental issues." But the need for this became a little bit more clear in 1998 with Hurricane Mitch which devastated Central America—Honduras, in particular. One of the things that permitted an effective disaster response and kept a democratically stable government in Honduras is that there was a mitigation plan in place. We had actually done an exercise the year before in Honduras that gave us a scenario of a major hurricane coming through Central America. True, we still had interagency problems. But at least the government agencies, civilian and military, were all working together and exchanging business cards before the crisis, not during it. And that is key.

Looking back at all this with the wisdom of hindsight, there are some important lessons to be learned. Honduras is still, as you know, a democratic society today. It survived Hurricane Mitch despite its devastating economic impact. Contrast that with what happened in 1972 in Nicaragua. A major volcanic eruption devastated Managua, Nicaragua. It killed about 11,000 people. That environmental situation exacerbated the political situation. Seven years later, the government was overthrown, and the Sandanista

Regime took power. And, oh by the way, Daniel Ortega is back. He has actually, now, been elected as President again.

So environmental threats do have national security implications. I think our conference has done a good job discussing that. I think this panel has done an excellent job of explaining the practitioner's role. And I think that they agree, that, regardless of whether we can agree or disagree on global warming, the military still has the responsibility to respond to the challenges ahead. Thank you very much.

Discussion

Q: Military and civilian institutions need to coordinate more effectively when it comes to planning. Yet, U.S. civilian agencies do not have the culture, the resources, and the tradition of planning that the military does. Could each of you mention one quick pragmatic thing that we should be doing in the next 2, 3, or 5 years to better integrate and prepare for planning across military and civilian U.S. federal agencies?

Tussing: We are trying to establish a strategic cultural mindset in the Department of Homeland Security. There are no less than 64 detailers sent from the Office of the Assistant Secretary of Defense for Homeland Defense and American Security Affairs to develop an educational apparatus. There is a document called the *Department of Defense Strategy for Homeland Defense and Civil Support*. It speaks of three pillars of operation: to lead, to support, and to enable. There are some things that only the military can do. Then it leads. There are some things that it can do in support of civil authorities. Then it supports. Sometimes it builds partnerships inside and outside of the United States and helps others acquire the capabilities they need to deal with problems. Then it enables. This, too, is very important.

Nash: On the State Department's side, the Office of the Coordinator for Reconstruction Stability has initiated a planning effort. The services, for the most part, have been reasonably good in supporting them and in sending officers to the State Department. Also, the Joint Forces Command has been very helpful in developing an ethos, if you will, of both planning and training—two issues the military is strong on and most civilian agencies are not. And, lastly, the military must learn to lead from behind when it works with civilian agencies at home, just as it does when it works with foreign militaries abroad. That means giving credit and visibility to the local authorities.

Johnson: Historically, the U.S. military has been one of the worst long-range planning agencies in the country. For three wars—the Spanish American War, World War I, and World War II—we had to call on big business to help us figure out how to do our job. The existing organization (then 250,000 people, more or less) was nowhere near comparable in size to the large corporations that operate in the international arena today. They say they do not have the manpower to spare for training. Philosophically, I am opposed to a larger government. But under the present circumstances, there are a lot of government agencies that need to get bigger so they can send—or believe that they can send—people off to do training. Training on the job is really cool, so long as you get the training before the crisis hits. We in the military have a very structured approach to training, and sometimes that's not good enough. We need to look at a variety of alternative paths and continue to talk to people, including big business.

Nash: I was surprised that Colin Powell, when he was Secretary of State, was not able to make more progress on the professional development of Foreign Service Officers. We need to establish a system that would facilitate that. That is, I think, one of the things that should be on our agenda.

Q: When you are engaged in operations, you find yourself in a place that is culturally constrained and has customs and a language that, in the post-Cold War era, few people may know. How ready are the armed forces to deal with this?

Nash: We can be very well-prepared if we think things through. The problem is not always well-understood at the senior level. But the Armed Forces of the United States have had a lot of experience operating in foreign lands. You learn over time that if you treat people with dignity and respect and recognize their worth, then you can adapt quite easily. You can read and listen to tapes and get a simple understanding of a different culture. You will also gain respect if you live by your own values. You have no doubt heard of the DIME—Diplomatic, Information, Military, and Economic elements of power. Well, I would argue that values and behavior are also an element of national power.

Johnson: Any soldier who wants to, can take any language he wants to through the Rosetta Stone process. It will at least let you begin a civil discussion. It is free online to you, as a soldier, which, given how costly this is, tells you something of how seriously we take this language business. The next step, of course, is to reward soldiers for demonstrated proficiency in the languages you are most interested in. The Army publishes its top ten list. And soldiers going into a theater are expected to be at least minimally conversant with the language.

Nash: George Blanchard, when he was Commander in Chief, U.S. Army Europe in 1975, ordered that all 200,000 plus soldiers learn to speak German. Many battalion commanders and above got sent out to Monterey and were ordered to include a German section in all change of command speeches. And so, you know, it had an overwhelming impact on the relations with the German people.

Audience: I teach at the Air Commander Staff College right now. We are teaching four major languages to students—Mandarin Chinese, French, Spanish, and Arabic. All the students are tested when they initially arrive, and then they are assigned to learn the most difficult language based on their test scores. So during the year, they go through the Rosetta Stone process and learn a foreign language. They are also exposed to 88 different international officers from 76 different countries, as are the students at the Army and Navy schools. So they are exposed to other cultures from their overseas assignments as well as their interaction during school.

Q: How do you define environmental security? And is there a common definition within the commands, and what is it?

Tussing: I do not believe there is a common definition. As I see it, the purpose of environmental security is to make sure people have protection and provision. The government, through its military, can make people's lives better. Environmental security is an extension of this idea.

Nash: Security has been a principle of war for a long time. A commander is required to consider how to "secure" his force as he conducts his military operation. That has been translated into force protection and, often times, becomes all too consuming. Public "security" is a mission, as opposed to a principle of operation. In Iraq after the surge, the order was given to "secure" the public. That had more impact than the number of soldiers sent. You can take that same concept and apply it to any number of functional areas to include environmental security.

Tussing: "Security" is a very important word. Note the distinction between the words "security" and "defense." There are very few people in the United States who have any question about who the guys are who are in charge of defense. But, "security" is a concerted national effort. And it does not begin and end with the guys in green.

Q: I was surprised not to hear much reference to the National Guard and Reserve components.

Tussing: The National Guard has what amounts to a state sponsorship program for individual countries in the world. And they have built up a working relationship with others for the purpose of sharing experiences. For instance, the state of Arkansas and the country of Kazakhstan have probably more ties than one would really want to get into. Remember, though, that for the preponderance of their existence, the National Guard belonged to the state governors. And, in spite of recent history, they will generally not have a national mission, let alone an overseas mission. But you can be sure that particularly with the growing concerns about the potential disasters, not natural but man made, the governors are keeping a closer and closer watch over, not just how their guard is responding, but how their guard is interacting with regional guard organizations within the United States.

Audience: About 4 years ago in the Air Force Reserve, we set up and identified reservists. These were composed of academics, people in business, etc., who had lived oversees for a long time and had a lot of good contacts. We drew up a list of these so we could call on them if needed.

CHAPTER 8

KEYNOTE AND SUMMATION

This final chapter brings together two talks, given at different points in the conference. The first essay is by our keynote speaker, General Paul Kern. In it, he recounts his own growing understanding of the serious implications of climate change and urges us all to address the problem. Dr. Richard Weitz provides a summary of the Triangle Institute for Security Studies (TISS)/U.S. Army War College (USAWC) conference, drawing attention to the areas of consensus and controversy. He concludes by comparing the findings of recent study undertaken by the Center for Strategic and International Studies.

Keynote Address

General Paul J. Kern, USA-Ret.
The Cohen Group

Introduction by Dr. Douglas V. Johnson II.

It is an honor to introduce General Paul Kern. General Kern is a graduate of the military academy, as am I. General Kern served two tours in Vietnam, as did I. General Kern got a master's degree from the University of Michigan, as did I.

General Kern then went on to movies and then science and technology and other related fields, and at one point, the Army, in a stroke of wisdom, uncharacteristic perhaps, said, "General Kern, you need to take command of this division with which we are going to conduct a very high technology experiment." So they gave him command of the 4th Infantry Division (Mechanized), and they said to him, "Make it all connected and make it all connect correctly, and work synergistically, every single piece in harmony with every other piece." And he laid the groundwork for that thing, and ultimately produced something that was absolutely spectacular, just in time for the Army to decide they didn't need divisions; but the lessons we learned from that organization process, that training process, that equipping and digitization process has moved us decades ahead in increasing our capabilities with the present force. General Kern ended his career, for all practical purposes as far as I'm concerned, with command of the Army Materiel Command. That is the organization that equips the Army; that provides it with reliable equipment in the proper quantities and qualities; and does so based on General Kern's experience with the 4th Infantry Division, so organized and digitized that all the pieces connected—at least most of the time.

The Army unfortunately lost his services, and he is now consulting on environmental issues for the Center for Naval Analyses. But our paths did cross one more time before he retired, and that was when I was teaching a transformation seminar out in our war college. A colonel came up to talk to the class and was diverted immediately after that. General Kern sat down at my seminar table with about 22 guys and gals, and gave us the frankest evaluation of the state of the Army and its equipping and equipment process that any of us have ever had. He gave us a view of the potential for the future which absolutely took our breath away. This is a man who knows how things work. Maybe he will get the environment to work, also. Sir?

Address by General Paul J. Kern.

Thanks. That was a nice introduction, and I appreciate it. I enjoy being back here at Duke. I think if we can walk away tonight with a concept of how all the pieces fit together, we will have been successful; however, I fear that to do that we might have to stay here for another couple of months. This is a very challenging topic that you have picked, and it has been a very challenging seminar. The issue is one I have been very much engaged in. I recently have been part of a military assessment board that the Center for Naval Analyses put together on global climate change. I have to give great credit to Sherry Goodman, who put together a series of seminars for about 12 flag officers of the Army, Navy, Air Force, and Marine Corps, and said, "Come out with a single paper." Now think about that. You take a noncontroversial topic like global climate change, make a military assessment of what is going to happen as a result, and ask all four services to come up with the same answer. That's almost an impossible task. But we are just about there, and within a couple of weeks, you should be reading about what we have all decided.

Burning Refineries and Fast Cars.

Now I will try to do that for you. You heard a little bit about my history from Doug Johnson, but maybe I need to put it in a little bit more context. I grew up as a young kid in New Jersey, and I used to, on a summer morning, go to the top of the hill where they had an ice cream restaurant, Gruning's, located on what we called First Mountain. Now First Mountain is about 300 feet high, but it overlooks Jersey City, New Jersey, and off into the New York Harbor. And for fun and games, we used to watch the refineries burn. Every year, it was very predictable that we would have a major fire in one of the refineries—you only need to look at all the oil tanks that sit down there to figure that one out. And the tanks could burn for quite a while, and we didn't think anything about that, other than that it was a big fire, and somebody was going to have to go put it out.

We didn't think much about what this meant from a national security perspective. We didn't wonder if someone had intentionally gone to set that fire. Nor did we think much in terms of what it was really doing to the environment. At that time, all the cars we drove were as much hot rods as we could figure out how to make them from an old V8, and the last thing that we considered was what was coming out of the tailpipe. What we considered was how much gas you could get through that carburetor and if you could super charge it, because then you could get a little bit more . . . and don't get caught by the cops. And so that was the environment that I grew up in.

Vietnam, Agent Orange, and Army Missions.

I went off to the Military Academy—Bill Nash and I were there at the same time—and I can guarantee you that in 4 years of studying, the thoughts about global climate change and the environment never crossed our desks. Much like the young graduates of all of our ROTCs, academies, and officer candidate schools today, we were focused on what was going to happen after graduation. And, for us, that was going to Vietnam. In Vietnam, we dumped a lot of Agent Orange to get rid of all that foliage. We used napalm to burn off jungle, and we didn't think, again, too much about what that meant with respect to the environment. That was a military operation. Napalm was one the tools that we had with which to conduct those operations, and so we used it. That was how we thought for quite a while.

Well, after spending some time in a Navy hospital, I went on another tour to Vietnam. I looked around at my friends—McCaffrey and Donovan (who were badly wounded), and looking at myself in the mirror as another Irish descendant—I said, "Geez, I wonder if anybody's figured out that, though my name is German, half my family came from Ireland (some think from a family of horse thieves that were running around Ireland before they came to the United States), and maybe the Irish's luck is running out." And so when I was finished with my second tour with the Army in Vietnam and was asked, "Would you like to go to graduate school," I said, "Yeah, that sounds like a good deal."

Michigan: First Encounters with the Environmental Protection Agency.

So I went off to the University of Michigan and had a great time. The Michigan/Ohio State football rivalry was, at that time, in its heyday and was absolutely one of the greatest rivalries of all time. Bo Schembechler and Woody Hayes were providing as much entertainment on the field as were the football teams.

It was at Michigan that I got into this business of looking at the Environmental Protection Agency (EPA) which was actually being established at that time. In fact, the Agency put its first test offices within Ann Arbor. I was looking at transportation through studies in civil and mechanical engineering. And I took part one summer in a study of combustion which asked, "How do you clean up what is coming out of that tailpipe?" This was a new idea, and I had a really good friend at that time who had been a Military Police (MP) Sergeant. I was an Armor Captain in a cavalry unit. We had great stories about what the Army was doing, and we were both suffering terribly in these advanced differential equation courses and a few other courses that we had to take from the math department for the summer. Well, he went off to work at the General Motors (GM) Tech Center, and I had another year to finish up. As we did that, we would look more at those studies. And at that time, the country was talking about the new rules that were being passed to govern fuel economy—the Corporate Average Fuel Economy (CAFE) rules. At the same time, as discussed earlier today, we were trying to figure out how you could improve fuel economy and control emissions—how you were going to regulate the emissions coming out of the tailpipe.

Well, as a bunch of mechanical engineering students, we said, "Man, that's going in the opposite direction. How are we going to do this? You can't both regulate emissions coming out of the tailpipe and improve on fuel economy at the same time." That was the thinking that was going on back then. We further thought that you certainly can't do it and be competitive with Toyota. At least that was how the automotive companies felt—they were feeling the pressure from overseas sales. As far as they were concerned, every penny that you spent on a car to improve its capability was a penny less that you put into the corporate revenue.

Well, my buddy who was the MP Sergeant went off to work for an automotive company. A couple of years later, after both of us had finished with work, he said, "If the government had never passed that law about fuel economy and emissions coming out of the tailpipe, the automotive companies would never have done anything." And so it was a policy set of rules that was not very popular politically, nor very popular on Wall Street, that said that you had to do something.

Now shortly after that, we all sat around and went in odd and even days to the gas stations, and we sat around and watched what really happened in terms of fuel economy. I was a graduate student at the time and started forming an opinion, and I wrote a paper that was wrong though, fortunately for me, my professor agreed with me. I looked at the energy and the environment as a whole, and I said, "You can't do both. You've got to make some choices." And, equally fortunately, there were some smarter people around who said, "No, you have to do both," and sent us down a path to accomplish controlling emissions *and* improving fuel economy. Now that was almost 40 years ago, and we are at a position again today where we are going to have to make some hard choices.

Changing Priorities.

A couple of years ago, Sir Michael Jackson, a good friend of mine and of Bill Nash's, was giving a lecture in London on what the future challenges would be in defense. I had to get up first. Now I had recently listened to a lecture that had been given by a Rice University professor, a Nobel Prize winner, Rick Smalley. He has since passed away. Some of you may know him. Carbon nanotubes were really his forte, and buckyball was his real masterpiece. Anyway, I had heard him give a lecture to a bunch of grade school kids. We had started a contest to try to get some interest in math, science, and engineering. And in that lecture, he said, "The country has 10 problems to face in the future. Number one is going to be energy." He went on to list the next problems, and the top four were energy, water, the environment, and food. And I thought about that, and about what I had said some years ago, and about the positions that we were all in. And I stood up in front of this audience of Brits and said, "Our top three problems are energy, the environment, and water," and depending upon where you live in the world, you can change that order, but if you live around a desert, and you have a lot of saltwater, it takes energy for desalination to produce fresh water that you can drink.

If you live in a desert region in the western part of this country and you want to irrigate, you will need energy. Some people think that California naturally has lots of water running through it to produce all those vegetables and fruits. Not true. It takes energy to pump that water across those mountains and into the valleys so you can irrigate. So again, you have this balance between energy, and water, and your environment.

Now this military assessment board that we looked at started putting these pieces together, and we said, "Is there really something here? Is there a military security relationship between energy and the environment?"

Now we didn't add energy into it initially. We looked at it from the environmental perspectives, and you will hear a little bit more from Tom Morehouse about the energy perspectives tomorrow. We started by saying, "Well, we are not sure." We listened to lots of people. We listened to some of the most alarming theories about falling off the cliff, very precipitous changes, and considered some of the things that could happen. We looked at things like what happens when the permafrost melts in the tundra regions and asked, "What do the models predict about that?" The answer is that we don't know, but we do know that inside that permafrost is captured some frozen methane—if you are a Brit, it is **me**-thane—and we do know that that is going to add to the greenhouse gases without our doing anything. We do know that the earth is going through fairly predictable cycles of climate change. Now the good thing about those cycles is they are longer than your and my lifetimes, so you and I will not have to deal with it while we are living; but we do know that the earth will have to deal with it.

The interesting thing when we went over to the United Kingdom was that they put both of those models together. They looked at human input, greenhouse gases, as well as the history of the earth's orbit, and figured out that together the models predict that the environment would get worse, though the models were not accurate enough to predict exactly when. So we started thinking through that, yes, there really are going to be some issues here that we probably ought to start thinking about so that we will be in a position to do something if we have to react.

Business Enters the Picture.

Well, that was step one; but perhaps to me what was the most intriguing and informative (and you heard a little bit about it today, too) was when the business world stepped in and said, "We want to play in this issue." The number one industry that caught my attention was the insurance business. Now what does the insurance business have to do with climatic change? The answer is that they pay the bills. They have been paying for these extreme weather events that we have been seeing. We have watched insurance companies go under, go bankrupt, as a result of not having the resources to meet the cost of multiple calamities.

Take places that are in the coastal regions, and especially the Gulf coast of the United States, whether it is Florida, or Louisiana, or Mississippi. You cannot get insurance there now. So people are feeling the economic impact of the storms in a way that had not been predicted before. I ask you, is that global climate change, or is that a local weather change? Is that part of the historical trend or is that something that is a result of greenhouse gases that we are adding? We debated that for a while. We listened to a lot of different opinions, not all in agreement. However, the answer you come back with after listening to this kind of discussion for a while, is that we are not really sure that we can pinpoint exactly what is going on here, but still we ought to be prepared, because something **is** changing.

Now as a young kid, I made a lot of money shoveling snow and watched a lot of hurricanes bring the bay and the ocean together in New Jersey. That stopped for a while, and now we see it coming back again. So, perhaps, the storms aren't just due to all the things that we are adding with greenhouse gases. Maybe there is something in the cycle that causes the extreme weather. But when it came down to a military assessment, we kept going back and saying, "Something is happening. We can be very certain that with all the scientific data and with all the climatic data that has been put together, there is something that we ought to be prepared to do."

So we started looking around at the different combatant commands. We looked at the geographic areas and looked at the pieces. And at that point, we pretty much recognized that, as you all discussed today, one of the things that we can do is to begin an engagement process with people and help them deal with whatever it is, whether it is a tsunami, a hurricane, an earthquake, or some type of a natural disaster—and no matter what was the cause of that natural disaster. If you are caught up in one of these disasters, it doesn't make any difference about the cause. Who you see coming to help you does make a difference. We also saw that there was pretty good evidence that both the methane, which was mostly naturally occurring, and the CO^2 concentrations, which were man-made, were contributing to this, and so we started down that path in trying to look at it.

Planning and Resource Allocation.

At this point, we got the admirals and generals together and said, "Okay, let's come to some agreement." Well, we knew a fair amount about what happens in military organizations, but we did not have the scientific background to really be sure whether climate change was man-made, a result of natural events, or some combination. But the bottom line was, if you are a military planner and if you are building organizations and putting resources against them, you ought to be prepared for the things that are out there. If those things out there are going to make the situation worse, then you ought to figure out how you are going to deal with them. You have to apply resources to deal with problems you see on the horizon. That is really what we call troop to task business. You carry out a military assessment, establish a mission, and allocate resources to it. And it was fairly clear that if we did not take climate change into consideration, there was going to be some multiplying effect.

To give you an example, Bill Nash and I watched the Sava River go from a little trickle to become a major problem. It required all new bridging assets to get into the Balkans. Well, I can tell you when we started looking about going into Iraq in the 2003 time frame, we noticed that there were a few rivers around

Baghdad. We looked at all the bridging assets and at all the flooding possibilities. Floods could be man-made or they could be natural, but one way or another we were going to have to deal with them from a military aspect. Similarly, people who look at the Arctic ice sheets look at them from a military perspective. We used to send the submarines there all the time. We do not do much of that anymore. We used to send the icebreakers up there. We don't have many icebreakers anymore. Making these shifts takes resources, and those resources do not materialize overnight. You have to think through how to conduct your operations. Experience tells you, "You have to plan for it. Don't get surprised by it." You may not have the right plan, but when you are looking at a national strategy that says that the environment and global climate change are something we ought to be prepared to do, you have to look at what are the second and third order questions about the kinds of resources that will be needed. It takes planning, and it takes prior planning to get those into the flow. That was one clear conclusion on which we could all agree.

We also agreed that, if you are going to deal with this problem, you need to recognize that there are lots of publics out there. There is a European public which has come to much more of a cohesive agreement than we have. There are lots of different American publics out there. And if you really want people to pay attention to something, you cannot afford to alienate any one group. If you do that, the argument becomes focused on a small microcosm of the problem. It is better to step back and say, "Geez, we really need to think about this seriously. It's energy, it's water, it's global climate change, it's extreme weather conditions. It's all of these things that we are going to have to deal with. How do we plan for it?" Don't let the issue become focused on a specific source of global climate change.

If you are looking at it from a military perspective, the important question is, "How are we going to deal with the problem?" How are we going to find a response mechanism, whether it is an engagement mechanism, or whether it is setting the example, or whether it is improving our use of energy and decreasing our greenhouse gas emissions at all of our installations. There are things that we can do to set an example, and there are things that we can do to be prepared. If the assessment is that there are severe climatic conditions in the future which will impact military plans, then we can let the scientific community argue the why and the how we will get to these conditions. Our task, in the military, is to step back and say, "What should we do to prepare? These are some of the conclusions that we have come to."

Economic Incentives.

One thing that you brought up time and again in the discussion today is the importance of economics. This is a lesson that I have learned out of all of this, out of watching how people solve problems in our world today. Not everybody is in 100 percent agreement, and so you are going to hear some continued debate here, but I think it has been a very healthy set of arguments. You are beginning to see that there are people who are investing money in trying to solve problems, whether it is Wal-Mart, General Electric, Virgin Airlines, or the insurance companies. They are saying, "It is time to step up and put some of our money into solving these problems, because if we don't, we might create a problem that is worse than we expected. We have, if nothing else, some years to learn how to become more efficient and to learn how to adapt to the changes. We need to invest in results which will keep us competitive." And so economic incentives play a big part.

I attended an Air Force sponsored energy symposium a couple of weeks ago, and one of the things we talked about was biofuels. Well, we have been talking about biofuels for a long time. Is it 5 percent ethanol or is it 85 percent ethanol? Do you go out and collect all the grease from McDonald's and pour it into your fuel tank and burn it? You can do that, and some people have done it fairly effectively. But questions remain. What is the right economic solution, and how do you provide the right economic incentives so that what comes out of the tailpipe and what goes in the front end to produce power are the most efficient and the least harmful to the environment? And another set of military assessments says, "Every dollar that we spend on petrol that does not come from a well inside the United States is a dollar that we give to a terrorist." And there is another set of economic incentives that says, "Well, maybe there are good reasons to not be so petroleum dependent."

If you live in Africa, you worry about emmigration caused by the expansion of the Sahara. The data that we saw said the desert grows about one mile in radius every year. What does that cause? Well, it causes people to go in search of more arable land, and they want to head north to Europe. This is yet another problem that our European friends have to deal with. So there are other issues involved in this whole business of the environment and how it is changing our world. But again, it is fairly clear to me that what you really get a return on is incentives. To go back to this Air Force conference; when they gave a real tax incentive to biodiesel, guess what people did? They started producing a lot of biodiesel fuel. The quantities increased almost proportionally to the tax writeoff.

A second thing happened, too. Whether you are talking about a 20 percent, or a 30 percent, or a 40 percent, or 100 percent biodiesel engine, you have to meet a number of other standards. Failing to meet the fuel standards begins to adversely affect the capabilities of our current engines. Now the blame game starts, "Well, it's because, of that biodiesel fuel." The answer is wrong. It's really because, if you look at it from a technical point of view, the fuel does not meet all the standards that we require of either a pure gasoline or a pure diesel fuel today. The fuel standards that we have created make sense. You cannot just throw them out the window, or you will be looking at the negative impacts and the incentives from a biased set of data. This will create positions for special interest groups based on erroneous data sets. The scientific and academic community must enlighten everyone on the facts and conclusions from accurate assessments. I encourage you to continue this debate and publicize your results. We need the unbiased facts.

What makes people change today in this world, particularly in this country, are economic incentives—we are a democracy, but we are also capitalists. What counts is how people see the return on their investment, and this has an impact on how people behave. And so the incentive process needs to do more than just appeal to people's view that we are improving the health of the world and that we are doing something for a greater global good. It also has to address the return on your investment.

The International Dimension.

Finally, there is the more international perspective. When you look at two great countries in this world as economic challenges to the United States, India and China, we can make a threat out of them, or we can make friends out of them. How we deal with that is going to depend very much on how we deal with the energy and the environmental issues.

They feel that they have every much of a right to pollute this earth as we did when we started the industrial age. One of the questions that we are going to have to deal with, again from a total economic and national security perspective, is if the Chinese are building a new power plant every week to 10 days—and it is a coal-fired power plant, and the winds blow from the west—is it in our best interest to help them clean up those power plants? We could charge them for it. We could partner with them. We could do lots of things. But there are two issues that come out of that. One is national security, and the other is economic security. And we ought to look at all these problems as much more of a global set of issues because we are all living together on this biosphere. There is only one biosphere here, and we somehow or another have to figure out how to share it whether we speak the same language or burn the same coal or want to fight over the same piece of rock. This is the earth we live on.

And so I think what I have learned over the last few years is that we all look at this from a different perspective, and we will doubtless argue about what is causing climate change for some time before we decide exactly what it is. Indeed, we may never know exactly what it is. But the problem is real. The science may be bigger than we can deal with, at least in our lifetimes, but we do know how to solve some of these problems. We do know that there are some impacts that are going to affect our military resources, and we need to plan for that right now. Our report will be ready, we hope, in a couple of weeks for publication, about the same time as some of these other issues will be coming around. We appreciate that you have all taken this on.

I have great affinity for the university systems and the Army research offices in this part of the country. Molly Broad and I had some long discussions when she ran the North Carolina University systems about how it was impacting the military and what we could do to improve it, so I appreciate that the War College and the universities here are getting together and addressing this issue. I think it is a pertinent issue for our generation. My kids like to look at me and say, "Dad, you've really screwed this one up. We've got the world at war, we've got the environment going to hell. And you guys aren't retiring. You are working longer, so you are taking all the jobs!"

I think there is more than enough for all of us to do, and I think there is a great challenge that you have in front of you right now. So again, I congratulate all of you for taking this on, and forming this partnership to address this issue. It is not easy, and it is not one that everybody is going to agree on the first time through. There will be a lot of contention, but that is a necessary part of making our lives better and making the world a better place to live in during the 21st century. With that, I will try to answer any questions you might have. I know that you all just had 8 hours of lectures and 2 hours of happy hour, and it is late at night.

Synopsis and Concluding Remarks

Richard Weitz
The Hudson Institute

Synopsis.

Areas of Agreement. In this weekend's discussion on the relationship between climate change and national security, many participants seemed to share a consensus on a number of points. For example, the attendees at the colloquium shared the perception that climate has been changing in the past and is changing now. A general scientific consensus also exists that managing the problem will require not just a reduction of greenhouse gas emissions to curb the effects of climate change, but also an increase in the use of alternative forms of energy.

In addition, it is now widely accepted that climate change can seriously threaten U.S. national security. Direct threats to the lives and property of Americans could come from natural disasters on the scale of Hurricane Katrina. (Although the precise connection between climate change and hurricanes remains an object of scientific debate, standard scientific models do predict global warming will lead to rising sea levels.) They could also arise from diseases and pandemics induced by climate change or from mass migrations within or into the United States of threatened populations as coastal regions flood and agricultural breadbaskets disappear. Climate change also threatens U.S. allies and the Americans living with them. For example, any climate-induced disasters in Japan or Europe could easily impact the U.S. troops and their dependents based in the region.

Another consensus belief expressed at the conference was that climate change could bring the need for many more humanitarian interventions led by the United States. The Cable News Network effect (highly visible during humanitarian emergencies), pressure from friendly governments and nongovernmental organizations, or other ethical or pragmatic considerations could lead U.S. policymakers to feel compelled to rescue people from climate-induced disasters.

In addition, the military can begin to take initiatives on its own to prevent climate change from degrading its operational readiness as well as to enhance the capabilities of its forces to undertake humanitarian missions. The military can also begin to adopt more energy-efficient practices and technologies, some of which may prove suitable for adoption by the civilian sectors of the economy.

Conference participants also agreed that the military alone could not manage the consequences of such a wide-ranging and multidimensional threat as climate change. Instead, they stressed the need for an interagency approach that would involve deep and sustained collaboration between the military and a range of civilian agencies—from the intelligence community to climate scientists.

Within the military, participants argued that the Department of Defense (DoD) should incorporate issues relvant to climate-related environmental changes into DoD planning processes, including those relating to unexpected contingencies as well as routine theater-engagement plans.

Areas of Disagreement. Despite the consensus on these general principles, the participants differed on many specific issues. Agreed definitions of what constitutes a "climate change" issue as opposed to one concerning related environmental, economic, or energy security issues are lacking. The boundaries between these terms, and their interrelationships, remain contested. One viewpoint present at the conference, however, was that consensus over these terms is irrelevant as long as we know the kinds of concrete physical phenomena that will ensue—and can reasonably anticipate their impact in the natural world and especially on human behavior.

Disagreements remain over the immediacy of the threat posed by climate change. Some scientists speak as if they are thinking in terms of decades. Under this timeline, the military will have the opportunity to be proactive. For example, it can take steps to increase its fuel efficiency and develop new technologies that may pay dividends only decades from now. Others describe the threat as more imminent. They therefore call for urgent preparatory measures that would influence the military's short-term planning processes and perhaps even current operations.

While the conference participants generally agreed that the military will only be one of many agencies to address the problem of climate change, this also presents some hurdles. The U.S. defense community could bring valuable resources and capabilities (e.g., intelligence and medical capabilities) to climate change efforts, especially since it is better resourced than many civilian agencies (which also encourages others to turn to it in national crises). Nevertheless, military involvement in this area will generate budgetary requirements and result in more resources flowing to the defense budget instead of to already under-resourced civilian agencies. If the U.S. Government turns to the military for every major security issue, the military's status as the only agency that can effectively "get things done" will be self-reinforcing.

Opinions also vary over the harmony between the challenge of dealing with the national security consequences of climate change and other goals. Some people maintain that addressing climate change issues from a national security perspective will help combat terrorism and improve U.S. ability to project soft power. In essence, they hope to experience a "virtuous cycle" where our efforts to limit climate change's negative effects create positive, ancillary benefits in terms of other U.S. national security objectives. Others worry about budgetary tradeoffs, leading to underfunding of other priorities, and other conflicts and negative spillovers.

The 2004 Asian tsunami and subsequent U.S. humanitarian relief operation does show the potentially positive consequences that can ensue from the application of limited U.S. military power to natural disasters. The successful operation helped improve U.S. public perceptions in a predominately Muslim-inhabited area. Yet, it remains unclear how easily the boon to U.S. soft power on that occasion can be replicated in other instances. In some cases, the affected population might prove less hospitable to a U.S. military presence, especially if the United States was seen as having contributed to global climate change through its past policies. The local government might fear that Washington would try to use the occasion to coerce it to change its policies toward the region or even, as might be the case in a near-term intervention in Iran, pursue regime change under the cloak of humanitarianism.

While the conference attendees generally agreed that the U.S. military must incorporate climate change into its assessment and planning processes, not all elements of the military appear equally ready or willing to make such a commitment. Some of the presentations showed that the regional combatant commanders are beginning to consider how to incorporate climate change into their own agendas. Nevertheless, climate change is not their main focus, nor do they have force personnel assigned solely to deal with climate change. In addition, combatant commanders tend to focus on the short term, in 2 to 4-year windows, as befits their responsibility.

The institutional bureaucracy in the Pentagon appears more hesitant to make long-term plans. Fighting climate change is not the reason why most people join the military. The same cultural and intellectual barriers to dealing with climate change within the armed services arise as they do in the cases of post-conflict reconstruction or state-building interventions. It is not the role most people first attribute to the military. The debate a decade ago over whether to set up a force for humanitarian or post-conflict reconstruction issues ended with the determination that separate assets should not be allocated for such purposes, given overall limitations on available manpower. Given today's increasing stress on the military with the wars in Afghanistan and Iraq, finding the resources to devote portions of the military to fighting climate change and its associated problems will be exponentially more difficult. It comes as no surprise, then, that there is no intramilitary consensus on the future role the U.S. armed forces must play in preparing for the national security implications of climate change, and whether or to what extent this should affect future force structure decisions.

Given these resource limitations, some conference participants underscored the value of collaborating with U.S. friends and allies to pool defense resources in the military's response to climate change. Many opportunities continue to present themselves in Europe, but the most interesting developments come from Russia and China. Despite the very complex military engagement process the United States has had with these two countries, climate change is one area that all parties may feel more comfortable dealing with and discussing.

Elsewhere in Asia, there are a series of dialogues taking place in which energy conservation and climate change are being addressed. For example, China, India, Japan, South Korea, and the United States are all participants in a five-party energy dialogue. In addition, the recent Asia-Pacific Partnership on Clean Development and Climate (AP6) with Australia, China, India, Japan, South Korea, and the United States is looking at other means of energy conservation as a supplement or a substitute for the Kyoto Process. As another example of opportunities coming out of Asia, in January 2007 the 2nd Annual East Asia Summit produced a declaration on East Asian energy security that made climate change and energy conservation an important aspect of their definition of security. As well, despite strained relations between Japan and Russia, and Japan and China, they are discussing possible energy cooperation, which could eventually spill over into a discussion of climate change issues.

Center for Strategic and International Studies Report.

A study currently taking place at the Center for Strategic and International Studies (CSIS) provides an interesting addendum to the climate change-national security discussion that took place at this Strategic Studies Institute/TISS conference.[1] The

CSIS project, which will culminate in a book, brings together perspectives from two traditionally separate expert communities: scientists of climate change and social science and policymakers/political scientists. The paper by Jay Gulledge, a scientist at the Pew Center for Climate Change, describes the physical possibilities of each of three climate change scenarios. His climate change scenarios provide the foundation for three other papers, each of which explores one of his scenarios and examines their potential national security consequences.

The least severe scenario reasonably can be thought of as the most likely outcome since it is based on a mid-range projection of the Intergovernmental Panel on Climate Change (IPCC). The two authors, John Podesta and Peter Ogden, consider the possible political, economic, and military consequences over the next 30 years, should major elements of the A1B greenhouse gas emission scenario of the IPCC Fourth Assessment Report occur.

The changes under this scenario are described as "expected" because many of these forecasts may have become inevitable as a result of past human economic activities that have already dramatically increased the quantity of greenhouse gases in the atmosphere. In addition, there is little indication that either a technological fix or a major international agreement will reduce anthropogenic carbon emissions significantly over the next 3 decades.

Balancing this pessimism, the scenario assumes that climate change does not trigger any significant feedback loops—physical tipping points that, once set in motion, become hard to stop or predict due to their potentially discontinuous or chaotic nature. Such catalysts—for instance, a decline in surface snow cover due to global warming, which reduces the earth's capacity to reflect light and thereby results in more warming—could magnify the effects of climate change substantially. Finally, the authors underscore the importance of local political, social, and economic factors in determining the geopolitical effects of climate change.

Prudent security planners always consider worst-case scenarios in determining policies, and this volume is no exception. The dangers of underestimating the degree of climate change—and its effects on the earth—lead the authors to posit scenarios that assume the scientific literature may systematically minimize the extent of global temperature rise and its effects on the earth's physical and natural properties. Like other scientists, most climatologists tend to be conservative in assessing phenomena they do not thoroughly understand, especially on politically controversial subjects such as the causes and consequences of climate change. In addition, the IPCC report-writing process requires that participating scientists reach a broad consensus on causal statements, at least in terms of a range of probabilities, in order for these statements to be included in a report.

Partially for these reasons, some climate scientists suspect that IPCC projections may systematically underestimate future climate change. For instance, they note that the models used to project future warming either omit or do not fully account for certain potentially important positive feedbacks that could amplify warming (e.g., release of greenhouse gases from thawing permafrost, reduced ocean and terrestrial CO^2 removal from the atmosphere, etc.). In addition, there is some evidence (e.g., changes in global ice cover, rates of sea level rise, tropical storm patterns) that such feedbacks may already be developing in response to the present warming trend.

These considerations lead us to develop a second, more severe 30-year scenario that assumes IPCC projections underestimate the rate of change by twofold. In this world, described by Leon Fuerth, average global surface temperatures rise to 2.5-3°C above 1990 levels over the next 3 decades. During this 30-year period, dynamic changes in polar ice sheets accelerate rapidly, resulting in a 45-60cm rise of average global sea levels. In addition, the availability of fresh water decreases substantially in the most affected regions at lower latitudes (dry tropics and subtropics), affecting 1-2 billion people worldwide. Agriculture becomes essentially nonviable in the dry subtropics, where irrigation becomes exceptionally difficult because of dwindling water supplies. Desertification significantly expands the extent of arid regions in the low latitudes, taking previously marginally productive crop lands out of production. The North Atlantic overturning circulation slows considerably, with consequences for marine ecosystem productivity and fisheries. Crop yields decline significantly in the fertile river deltas because of rising sea levels and damage from increased storm surges.

The CSIS study teams recognize that climate change could lead to many unanticipated developments, but it assumes that abrupt, large-scale climate events are unlikely to occur during the next 3 decades. To explore the foreign-policy consequences of such developments, the third scenario projects the continued development of the second scenario for the next 100 years. By hypothesizing that rapid global warming will persist throughout the 21st century, we can consider the effects of a much more rapid loss of polar ice, a sea level rise of more than one meter, a dramatic slowing or shutdown of the North Atlantic overturning circulation, and massive die-offs of wet tropical forests in the Amazon basin. Assessing the effects of abrupt climate change is enlightening because, while communities often find it harder to manage rapid change than gradual transformations, this pattern is not universal. Sometime abrupt shocks galvanize effective responses whereas long-term trends can become lost in the background of more pressing matters.

The 100-year worst case scenario described in the chapter by Jim Woolsey is truly one of "cataclysmic climate change." It includes direct effects (e.g., rising sea levels), secondary effects (e.g., mass migrations), and tertiary effects (e.g., disruption of global oil shipments) whose diverse interactions generate a cascading wave of international security threats. The author posits two positive feedback loops—"tipping points"—the melting of tundra in Siberia and the melting of the Western Antarctic ice cap. Trapped within the Siberian tundra are roughly 500 billion tons of carbon in the form of methane, which is 20 times more potent as a greenhouse gas than CO^2. The release of this methane could provide a substantial catalyst to the pace of climate change. Furthermore, the melting of the Western Antarctic ice cap could cause sea levels to rise many meters very rapidly.

In terms of physical impact, this sea-level rise could transcend all others in the coming century. The sudden and steady melting of ice sheets around the world could lead to some 25 meters in sea-level rise and the end of civilization as we know it. Even with only a 2m rise, areas across the globe would be inundated, and national governments would find it difficult to address public policy issues beyond those promoting their own salvation. For example, the U.S. Government would face unprecedented levels of immigration. In addition, the U.S. military would find its reach severely reduced by new logistical constraints and the need to respond to many missions near U.S. borders. Many domestic and overseas ports and harbors would become useless.

According to the author, climate change is a "malignant" problem derived from nature, where seemingly insignificant behavior can greatly increase the chance of a "metastasis" in the system. Terrorism, unlike climate change, is a "malevolent" problem, driven by evil intentions. Yet, the world cannot ignore either problem since climate change exacerbates security concerns. The world's energy infrastructure remains highly vulnerable to the effects of climate change and terrorist attack. For example, sea level rise and chaotic weather patterns might interrupt oil production. In addition, attacks by terrorist groups might cause oil supply interruptions in exporting states such as Saudi Arabia. Finally, terrorists might attack the U.S. electricity infrastructure with a physical attack or one involving electromagnetic pulse (EMP). An EMP attack could cause unprecedented cascading failures of the U.S. electric power infrastructure and cripple the U.S. telecommunications, financial, agricultural, and commercial infrastructures.

In addition to these three scenario chapters and scientific background, two additional analytic chapters help form the book. In one chapter, Georgetown University historian John MacNeill surveys past instances of major global environmental transformations, and how communities and international politics were shaped by these events. This look at historical lessons and their limits covers a wide range of phenomena with persistent consequences, including disease pandemics, natural disasters, and resource shortages. These events may provide insight into how today's and future societies may respond to global climate change on a variety of levels. In the last chapter, the policies and perceptions of the major powers towards climate change will be analyzed, including those of Europe, Russia, India, China, and the United States.

With an abundance of new studies on the climate change-national security nexus coming to fruition, the areas of consensus and contention discussed at this colloquium will continue to be topics ripe for reexamination and further debate in the months and years to come.

ENDNOTES - Weitz

1. Kurt M. Campbell, Jay Gulledge, J.R. McNeill, John Podesta, Peter Ogden, Leon Fuerth, R. James Woolsey, Alexander T.J. Lennon, Julianne Smith, Richard Weitz, and Derek Mix, *The Age of Consequences: The Foreign Policy and National Security Implications of Global Climate Change,* Washington, DC: Center for Strategic and International Studies, November 2007, available at *www.cnas.org/en/cms/?1278*.

General Wrap Up

(Comments from the Audience)

Comment Speaker 1: My name is Elizabeth Leahy, and I work at Population Action International, which is a nongovernmental organization (NGO) in Washington, DC. Based on our name, particularly the first word, I think you can guess what our issue is. We do research on global demographic trends, and we advocate increased funding for international family planning and reproductive health programs. Amy Cohen, my colleague and boss, and I came to this conference because we believe that population and demographics are very important to all the issues that we discussed over the past 2 days. Often when we go to interdisciplinary conferences or meetings of this nature, we have to spend a lot of time just convincing our colleagues from other fields that population and demographics are relevant and matter. I was gratified, and maybe a little bit surprised, that we didn't have to do that so much this time. Many of the presenters have discussed how countries' national security, the impact of climate change in specific countries and regions, levels of effective governments, levels of development, are all affected by demographics. I can see that the presenters here are considering adapting more holistic definitions of national security than I heard sometimes in the past. Population is a dynamic issue, and many of you seem to recognize that.

Maybe a bigger problem, unfortunately, is that many decisionmakers do not realize that controlling population can help solve many global problems. The United States has consistently been the largest bilateral funder of international family planning programs.

For the fiscal year 2008 budget, it has been proposed that those programs take a 25 percent cut. Part of the rationale for that is that these programs have been successful over decades in many developing countries; and they have been. People are starting to talk more about population aging these days, which is certainly a genuine concern. However, we still have a world where 55 percent of the world's population lives in countries that are facing sustained population growth over time. Our current population of 6.5 billion is projected to hit 9 billion by 2050. I will not recite the whole litany of statistics. Sometimes we are asked, "What can the military do?" And in the past, we talked about the military's role in post-conflict situations including humanitarian reconstruction, planning, and threat assessment. I learned a lot over these past 2 days about many of the noncombat roles of the military. Many of the speakers have encouraged military integration with other sectors, international organizations, other agencies, and NGOs. I will just conclude by saying that we are here to second that recommendation, and I hope that we will continue to work together to develop further opportunities for collaboration between our sectors.

Comment Speaker 2: Richard Weitz's idea of writing a book doing some analysis of scenarios strikes me as enormously useful. It is hard to do that in a conference, but maybe at least a rough scenario might have kept us a little bit better on-track. We did waffle around in trying to figure out what national security was, and that range expanded uncomfortably for me by the time we were done. As a result, we spent a lot more time talking about the U.S. military and what it could do in what basically amounted to disaster relief than in dealing with global climate change in a more strategic sense. As I said, disaster relief is a reactive thing that we do now, but addressing the implications of national security and global climate change is a strategic long-range process. Let's face it, it is hard to get your hands around something like that.

Comment Speaker 3: We hit-and-missed it throughout the discussion, but consumption was a core issue beyond just energy. We talked a little bit about overfishing, we talked a little bit about the United States being not only the largest producer of greenhouse gases, but also the largest user of oil, and things like that. However, when it comes to consumption across the board—and that relates to the population issue also—we did not stress enough the fact that, according to some assessments, we are consuming about two and a half planets more than what we've got. The Millennium Ecosystem Assessment ascertained that 60 percent of our ecosystems are stressed and are being overexploited right now. So all these different environmental ecosystems are under intense pressure right now, and then you throw in the additional stressor of climate change, another one of those potential tipping points. But at the core is consumption, being led by the developed countries.

Comment Speaker 4: Let me second all of those remarks, but especially those by Doug Johnson. We can talk a lot about disaster relief, but there are several problems. The first is that we might get stuck in places longer than we like. But the larger thing is that, as Doug says and as Tom Barnett and I worked out, we are dealing with vertical scenarios versus horizontal scenarios. This is a long-term process, with things happening along the way. Maybe there are incidents the military gets into, but in that larger scope of things, in the mitigation and the adaptation business, it is a long incremental business.

Comment Speaker 5: Clearly, the evidence is out there that, yes, global warming is happening, and, yes, we need to address it—but how do you get the people involved? It is easy in the Cold War to understand that one nuclear bomb could ruin your day. But how do you force this issue on the American people? Do we bring back Burt the Turtle, the fellow who warned us to duck and cover? I guess we don't know, or no one has put enough thought into finding out how best to do this.

Comment Speaker 6: This ties in with the issue of strategic communications raised by Hank Gaffney, among others. The audience that we are primarily talking about here has got to be the American people. Governance also gets into it, leadership gets into it, insight gets into it, and vision gets into it. We even have to get to the point where we educate the American people as opposed to alarming them. If we still have skeptics, then there is something wrong with the message that we have put out so far.

But I wanted to relate to the things said earlier about the military. The impression that a lot of people still have is that the military will only bow to pressure and and will pay attention to the issue of climate change only if you absolutely force it to. I would suggest to you that the kind of mindset that leads young men or women to join the U.S. military is the kind of mindset that will be open to the idea that they must protect their society by protecting their environment. What you have to do is convince the military that what you are saying is correct. It is easier to do this when your leadership also tells you that, to be sure. But once you get the military on board, you will find no better proponent of the cause. No one better follows the rules or sets a better example on how to make sacrifices for the greater good.

Comment Speaker 7: I just want to say that when we talk about spending more time engaged in long-term deployments, it isn't us who will be going. It is our kids. Just remember that. I have a son in Baghdad right now; I would just as soon he not be there.

Comment Speaker 8: You were talking about the strategic communication problem, and I would like to say that I come away from this conference really quite impressed and amazed to find that this conversation is taking place among the military establishment at all, and especially at this level of sophistication. Yes, there is a lot we still need to figure out. What exactly is the potential mission here? Exactly what kind of forces could we deploy? What kind of resources do we want to put in? All this still needs to be worked out.

But this conversation is much more advanced than I certainly came here expecting it was going be or hoped it could be. This is good news for two reasons: First, because you provide a message of national security and climate change, which has not been a major part of the broader public conversation at all. And second, because you represent a set of messengers that carry great weight and great credibility, with a lot of different audiences that most of the rest of the public doesn't ever hear from. And I think that you have got to understand how important the work that is being done here by all of you, and your colleagues that are not here, really could be. You open entrees into communicating and convincing people who would never listen to an environmentalist, or an academic scientist, or even, say, a religious person. I think this particular nexus is really important at this point in the evolution of the debate.

Comment Speaker 9: I would like to pick up on something that Richard Weitz said. We still need to figure out how all this fits in with the war on terror and all the other challenges. The difficulty is that those people who do not want to talk about climate change are going to try to distract us by talking about the war on terror, or China, or some other threat. If you list the threats facing us, depending on the nature of the conversation, climate change can come out fourth, eighth, or even 26th. The fact of the matter is that all these challenges are different from each other, and they are different from what we thought they were over the last 50-60 years. We cannot come up with a simple message—like we are dealing with Godless communism. And it is hard for us to articulate the nature of the threat. We need climate change zealots, I suppose, as we need zealots for all good issues. However, those who deal in national security don't get to eliminate any of them just because they are not of concern. They have to prioritize within the context of all the threats to the country, and we need to make sure we make that argument in such a way that it is high enough up that list to be useful.

Comment Speaker 10: Richard, I thought you did a good job of summarizing. However, I would like to comment on two concerns raised over the last day and a half that you did not mention. One is what I would call a minor concern, and the other one is more of a major concern.

The minor concern is this: I heard a number of people say, "We have the truth. We have the message. We understand, but we need to convince the leadership. How do we do that?" I would suggest to you that you **are** the leadership. You run the risk of falling into the pothole thinking you have to commit to somebody else who is in a leadership position. If you are not careful, if we are not careful, that can become a "that is not my job" syndrome, and we really need to watch that. Because I think there is more need for leadership here in addressing this issue than you perhaps are prepared to accept and recognize. That is the minor concern.

The second concern I also heard from a lot of speakers—and I tend to agree with this—is that the successful solutions here are going to require cooperation and coordination with a whole lot of different people from a whole lot of different cultures, backgrounds, and experiences, who have different agendas, goals, and purposes in life. And there was a legitimate mystery as to "how do we do that?" I think we all have experienced trying to work in those environments and in those cultures, and we have come away, most of the time, frustrated with our inability to achieve any sort of reasonable goal or success. That, to me, seems to be one of the underlying requirements of success in this area that I think is going to be a major hurdle. How do we get all these disparate folks moving in the same general direction?

Comment Speaker 11: Just an observation. The center that I direct works fairly closely with the military, and has for some time. It also works with other groups, including a group of business leaders in California, a couple of other parts of the government, and a good part of the environmental community. One of the things I noticed over the past decade is that the military is developing a critical mass of knowledge and expertise. This has given it clarity on these issues that these other groups do not have. I am starting to see the clearest discussions, the most informed discussions about bird flu, or environmental change, or climate change, often taking place in environments which have a large military presence.

Now, as Bert Tussing made clear, the military is there to support, not to lead, in a wide range of activities. In some instances this was very clear to me. I participated in the Mark IV PAC exercise to develop a plan for bird flu. It was clear that there was a lot of expertise at hand: these people were well-briefed on the issue. They understood the issue, they were prepared to discuss it, and they were very pragmatic about what they needed to do. The State Department was clearly confused about the issue, unsure about it, and very hesitant. I wonder if maybe we are getting to a point where the knowledge is shifting into this part of our governance structure in a way that is not going be that easy to transfer over.

So I think it is important that we do what we can to communicate to these other groups, "We are taking this seriously. We are no longer bogged down in debates over whether this, or what that. We are moving to the next generation of discussion." So while I am very impressed by the clarity that the military has brought to these issues, I find it rather disconcerting that other parts of our country have not achieved that level of clarity at this point, including elected officials.

Comment Speaker 12: This brings up a very important point. If you all can go back and carry this banner, then do so. You should be asking, "How do we train our government officials, particularly our strategic leadership?" And the answer is, outside of the five-sided wind tunnel, not very well. This is, I think, because a significant sector of the U.S. Government thinks that time spent on education is time wasted. It is very easy for the military to be coy from time to time and talk about the effort that we put towards strategy. After all, we have the people to do it and the people who have been trained to do it. And we talk about how other people ought to be taking up the educational issue so that they can rise from the tactical to operational to strategic mindset. However, we are living in a world that is growing more and more complex. And we cannot afford not to develop our leaders in every branch of our government. Not just our military. We must instill that notion.

Weitz: May I interrupt this just one second? Jim Hanson testified before Congress, I think it was Monday, and he listed a number of steps that we can take to mitigate global warming, and he concluded, "But the most important step you can take, gentlemen, is campaign finance reform."

Comment Speaker 13: Well, when we talk about leadership right now, the primary concern of the Prime Minister of Iraq is not climate change. We have a bunch of four stars who have very, very full plates, and tend to have very limited horizons as far as their crisis management is concerned, so it is difficult to get them to look 15, 20, 50, or 100 years down the road. That said, Al Gore did put forward to the U.S. Government a proposal to create a future-focused planning and evaluation body. I believe it exists in law, even though I think there are zero bodies occupying the place, and it is probably not funded.

Closing - Alex Roland.

On behalf of the Triangle Institute for Security Studies, we would like to thank our host, the U.S. Army War College, and our various sponsors—the Army Environmental Policy Institute, Creative Associates, the Center for Global Change Institute, Nicholas Institute for Environmental Policy Solutions, the Environmental Change and Securities Program, the Woodrow Wilson International Center for Scholars, and the Department of Environmental Sciences at the University of North Carolina at Chapel Hill. And thanks to our summarizer, Richard Weitz. We are adjourned.

ABOUT THE CONTRIBUTORS

JOHN T. ACKERMAN is an Assistant Professor of National Security Studies at the Air Command and Staff College (ACSC), Maxwell Air Force Base, Alabama. A retired lieutenant colonel of the U.S. Air Force, He is also the Research Course Director for the ACSC Department of Distance Learning. His research efforts have included exploration into the relationships between sustainability and security, the international relations implications of global climate change, and the national security ramifications of global warming. Besides his published dissertation on *Global Climate Change: Catalyst for International Relations Disequilibria* (2004), Dr. Ackerman has written a number of papers and reviews, most recently, "Climate Change, National Security and the Quadrennial Defense Review: Avoiding the Perfect Storm," *Strategic Studies Quarterly*, Spring 2008. Dr. Ackerman's doctorate in Political Science is from the University of Alabama.

JOSHUA BUSBY is an Assistant Professor at the Lyndon B. Johnson School of Public Affairs at the University of Texas, Austin. Prior to coming to UT, Dr. Busby was a research fellow at the Center for Globalization and Governance at Princeton's Woodrow Wilson School (2005-06), the Belfer Center for Science and International Affairs at Harvard's JFK School (2004-05), and the Foreign Policy Studies program at the Brookings Institution (2003-04). He has a long-standing research interest in climate change with articles on the topic appearing in *Current History*, among other publications. Dr. Busby is the author of a 2007 Council on Foreign Relations special report on climate change and national security. A scholarly piece, "Who Cares about the Weather? Climate Change and U.S. National Security," will be published in *Security Studies* in 2008. His doctorate is from Georgetown University.

KENT HUGHES BUTTS is Director of the National Security Issues Branch, Center for Strategic Leadership, U.S. Army War College. Dr. Butts has served as a strategic analyst in the Strategic Studies Institute, U.S. Army War College, and taught at the U.S. Military Academy. He has also been a John M. Olin Postdoctoral Fellow in National Security at the Center for International Affairs, Harvard University. He has authored and edited a number of books. In a 1998 work, *Environmental Security: A DoD Partnership for Peace,* he looked at how international environmental issues can lead to instability and conflict that threaten U.S. security interests and may result in the commitment of U.S. forces. More recently, he has written on how to combat terrorism. He holds a Ph.D. in Geography from the University of Washington and is also a graduate of the U.S. Army Command and General Staff College and the U.S. Army War College.

ROBERT W. CORELL is the Global Change Director of the Heinz Center. Before coming to The Heinz Center in 2006, he served as a Senior Policy Fellow at the Policy Program of the American Meteorological Society and an Affiliate of the Washington Advisory Group. Dr. Corell is engaged in research concerned with the sciences of global change and the interface between science and public policy. He co-chairs an international strategic planning group that is developing a strategy designed to harness science, technology, and innovation for sustainable development, serves as the Chair of the Arctic Climate Impact Assessment, counsels as

Senior Science Advisor to *ManyOne.Net*, and is Chair of the Board of the Digital Universe Foundation. An oceanographer and engineer by background and training, he holds degrees (including a doctorate) from Case Western Reserve University and MIT.

ROBERT H. DORFF is currently Research Professor of National Security Studies, Strategic Studies Institute, U.S. Army War College. Immediately prior to this, he was Senior Advisor for Democracy, Governance, and Civilian Military Relations for Creative Associates International, Inc., a globally-engaged, international consulting company. He worked extensively with their "Security and Development Initiative," focusing on strategic approaches and policies that bridge the civilian-military gap in addressing the challenges of post-conflict and transitioning societies. Dr. Dorff is the author or co-author of three books and numerous journal articles, many of which focus on U.S. national security strategy, democratization and failed states, and peace support operations. Professor Dorff holds a Ph.D. in Political Science from the University of North Carolina-Chapel Hill.

GREGORY M. DOUQUET (Colonel, U.S. Marine Corps) holds a Fellowship at the Royal College of Defence Studies in London where he is pursuing a dissertation on national security implications of climate change in the CENTCOM Area of Responsibility. Professionally, Colonel Douquet holds dual specializations in strategic planning and in aviation operations. Prior to his fellowship, he served as a planner for future operations in Iraq for the Marine Corps. He has served two tours of duty in Iraq and was involved in the evacuation of noncombatants from Liberia in 1989; planning the

1999 United Nations Peacekeeping mission in East Timor, Indonesia; and leading a detachment tasked with recovering the USS *Cole* in Aden, Yemen, in 2000. Colonel Douquet also served as a presidential helicopter pilot and White House Liaison Officer and completed a headquarters tour with U.S. Marine Corps, Plans, Programs, and Operations.

HENRY H. GAFFNEY, JR. is a senior analyst at the Center for Naval Analyses (CNA), which produced a major study of climate change and national security on April 18, 2007. He has been at CNA since 1990, specializing in broad studies of the evolving world security environment, to include the future of U.S. defense and the national security implications of energy and climate issues. Prior to joining the CNA Corporation, Dr. Gaffney served for 28 years in the Office of the U.S. Secretary of Defense. He worked for more than 12 years on NATO matters, for 2 years as Director of the Near East and South Asia office, and for most of the 1980s as the Director of Plans in the Defense Security Assistance Agency. He served as an officer in the U.S. Navy from 1956 to 1959. Dr. Gaffney received his doctorate from Columbia University.

DAVID GILMARTIN is Professor of History at North Carolina State University. He was a Fellow at the National Humanities Center, 2001-02. He teaches courses on Asian history, the modern history of India, the history of European imperialism, the history of Islam in the modern world, and 20th century world history. He is the author of *Empire and Islam: Punjab and the Making of Pakistan,* University of California Press, 1988; and *Beyond Turk and Hindu: Rethinking Religious Identities in Islamic South Asia,* edited with

Bruce B. Lawrence, University Press of Florida, 2000. His most recent research focuses on the history of water control in the Indus Basin region, the heartland of the modern state of Pakistan, and on the historical relationship between local community, the state, and the environment. He holds a Ph.D. from the University of California, Berkeley.

MICHAEL H. GLANTZ is Director of the new Center for Capacity Building at the National Center for Atmospheric Research (NCAR) in Boulder, Colorado. He is interested in how climate affects society and how society affects climate, especially in Africa. Recently, his research has focused on how to forecast possible societal responses to the regional impacts of climate change. He has taught at the University of Colorado, University of Pennsylvania, Lafayette College, and Swarthmore College. In 1983 he became the first (and still the only) social scientist to become an NCAR Senior Scientist. In March 1990 he received the prestigious "Global 500" award from the United Nations Environment Programme (UNEP). He is the author of many works, including *Climate Affairs: A Primer,* Island Press, 2003. He has a B.S. in Metallurgical Engineering and a Ph.D. in Political Science from the University of Pennsylvania.

KAREN LESLEY HULME is Senior Lecturer in the Law Department at the University of Essex, United Kingdom. She currently teaches courses on public international law, international law of armed conflict, law of evidence, and tort law. Her research interests are in public international law, the laws of armed conflict, international environmental law, and the law of evidence. Her dissertation examined the protection of the environment in times of armed conflict. She is the author of a number of articles and *War Torn Environment: Interpreting the Legal Threshold*, which was the winner of the 2004 American Society of International Law Lieber Prize. Her doctorate is from the University of Essex, United Kingdom.

DOUGLAS V. JOHNSON II has been with the Strategic Studies Institute, U.S. Army War College, since 1985, first as Strategic Research Analyst and then as Research Professor of National Security Affairs. His 30 years of service in the U.S. Army included two combat tours, a variety of troop and staff assignments, and instructor duty at the U.S. Military Academy and the School of Advanced Military Studies at Fort Leavenworth, Kansas. His areas of expertise include Army transformation; futures; military history (especially World War I and military doctrine); civil-military relations; Just War theory and practice; and media. He has edited, authored, and co-authored eight books on these topics. In 2001 he co-authored a book with Steven Metz on *Asymmetry and U.S. Military Strategy: Definition, Background, and Strategic Concepts*. He holds a Ph.D. in History from Temple University.

PAUL J. KERN (General, U.S. Army-Ret.) was the Commanding General, Army Materiel Command (2001-04) and is now Senior Counselor at the Cohen Group and advises the CNA Climate Change and Security Study. Prior to his command at AMC, he served as the military deputy to the Assistant Secretary of the Army for Acquisition, Logistics and Technology and was the senior military advisor to the Army Acquisition Executive and the Army Chief of Staff on all research, development, and acquisition programs and related issues. General Kern also served in the Office of the Secretary of Defense and in several field units. He had three combat tours, including two in Vietnam, and one in Operations DESERT SHIELD/DESERT STORM. He holds Masters' Degrees in both Civil and Mechanical Engineering from the University of Michigan, and he was a Senior Security Fellow at the John F. Kennedy School at Harvard University.

RICHARD J. KILROY, JR., is Professor of International Studies and Political Science at Virginia Military Institute in Lexington, Virginia. At the time of the conference, he was Assistant Director of Military Programs at East Carolina University. He retired from the U.S. Army in 2004, after 23 years of active duty service as a military intelligence and Latin American foreign area officer. He was also involved in standing up the U.S. Northern Command as the military's response to 9/11 and supporting the homeland defense mission. Dr. Kilroy is the editor of *Threats to Homeland Security: An All Hazards Perspective,* J. Wiley and Sons, 2007. Dr. Kilroy holds a Ph.D. in Foreign Affairs from the University of Virginia.

ANTHONY LEISEROWITZ is Director of the Yale Project on Climate Change and a Research Scientist at the School of Forestry and Environmental Studies at Yale University. He is also a principal investigator at the Center for Research on Environmental Decisions at Columbia University. He is a widely recognized expert on American and international public opinion on global warming, including public perception of climate change risks, support and opposition for climate policies, and willingness to make individual behavioral change. His research investigates the psychological, cultural, political, and geographic factors that drive public environmental perception and behavior. He has published over two dozen articles, chapters, and reports, including, "American risk perceptions: Is climate change dangerous?" *Risk Analysis*, Vol. 25, No. 6, 2005, pp. 1433-1442. He earned a Ph.D. in Environmental Science, Studies, and Policy from the University of Oregon.

RICHARD A. MATTHEW is Director of the Center for Unconventional Security Affairs and Associate Professor of International and Environmental Politics in the Schools of Social Ecology and Social Science at the University of California, Irvine. He is also a member of UNEP's Expert Advisory Group on Environment, Conflict and Peacebuilding; the Senior Fellow for Security at the International Institute for Sustainable Development (IISD); a member of the World Conservation Union's Commission on Environmental, Economic, and Social Policy; and a member of the Homeland Security Advisory Council (Region 1). His research explores ways in which conservation and sustainable development can be designed and implemented to reduce violence and insecurity in different settings. He is the author of many books, co-edited books, and articles including *Contested Grounds: Security and Conflict in the New Environmental Politics*, SUNY Press, 1999. His doctorate is from Princeton University.

TIMOTHY J. MCKEOWN is Professor of Political Science at the University of North Carolina, Chapel Hill. He has also taught at Carnegie-Mellon and Duke. Professor McKeown's current research investigates foreign aid programs and the practice of conditionality and soft influence strategies; government research and development subsidies as a trade-promoting strategy; industrial structure and foreign economic policy. Methodologically, he continues to work on developing the theory and methods of qualitative research, especially the systematic observation of archival material. He is the co-author with Leonard Lynn of *Organizing Business-Trade Associations in the U.S. and Japan*, American Enterprise Institute, 1988; and co-editor (with Dan Caldwell) of *Diplomacy, Force and Leadership: Essays in Honor of Alexander L. George*, Westview, 1993. He holds a doctorate from Stanford University.

E. THOMAS MOREHOUSE, JR., is an adjunct at the Institute for Defense Analyses (1994-present) and consults a number of companies and research institutions, such as the CNA Corporation. The confluence of climate, security, and energy issues has been a focus of his work for over 15 years. He was the principal author of two Defense Science Board reports that showed capability and cost improvements by reducing battlefield fuel burden. He also assessed the carbon benefits of Federal investments in a wide range of technologies to mitigate global climate change. His current work includes developing military responses to address the national security implications of predicted global climate changes caused by elevated atmospheric concentrations of greenhouse gases. He is the co-editor, with Philippe G. Le Prestre and John D. Reid, of *Protecting the Ozone Layer: Lessons, Models, and Prospects*, Kluwer, 1998. He holds an M.S. in Mechanical Engineering from Boston University.

WILLIAM L. NASH (Major General, U.S. Army-Ret.) has been with the Council on Foreign Relations since March 2001. A veteran of Vietnam, Operation DESERT STORM, and 34 years of Army service, he has extensive experience in peacekeeping operations both as the commander of the 1st Armored Division in Bosnia-Herzegovina and as a civil administrator for the United Nations in Kosovo. In addition to his duties at the Council on Foreign Relations, Major General Nash is an adjunct professor at Georgetown University, a visiting lecturer at Princeton University, and a military consultant to ABC News.

CAROLYN PUMPHREY has served as Program and Outreach Coordinator for the Triangle Institute for Security Studies since 2000 and also teaches history at North Carolina State University. From 1997-2000, she was a Post-Doctoral Fellow for the Triangle Institute for Security Studies, and between 1986 and 1992 she was an Assistant Professor of History at Spring Hill College (Mobile, Alabama). She has taught a wide range of classes dealing with war and peace in the Middle Ages and has edited three books, *Transnational Threats: Blending Law Enforcement and Military Strategies*, Strategic Studies Institute, U.S. Army War College, November 2000; *The Rise of China: Security Implications*, Strategic Studies Institute, U.S. Army War College, January 2002; and with Rye Schwartz-Barcott, *Armed Conflict in Africa*, Scarecrow Press, August 2003. She holds a doctorate from Duke University.

ANDREW PRICE-SMITH is Assistant Professor of Political Science at Colorado College and Director of the Project on Energy, Environment and Security at Colorado College. He held previous appointments at both the Earth Institute and School of International and Public Affairs of Columbia University, and the University of South Florida. Dr. Price-Smith's published works include *The Health of Nations: Infectious Disease, Environmental Change, and Their Effects on National Security and Development*, MIT, 2001; and *Contagion and Chaos*, MIT, forthcoming 2008. He has served as a consultant to the United States Institute of Peace, the United Nations Development Program, the World Bank, and the U.S. Department of Energy. He earned his Ph.D. in Political Science from the University of Toronto.

SIMON RICH is the former Chairman and CEO of Louis Dreyfus Natural Gas and President of Louis Dreyfus Holding Corporation; the Louis Dreyfus Group holds a global position in all areas of energy and agriculture including production, processing, and merchandising. He is also former chairman and emeritus member of the board of visitors of the Nicholas School of the Environment and Earth Sciences at Duke University, and teaches courses in Energy and Environment within the school. He is chairman of Environmental Defense in North Carolina, chairman of the Center for Environmental Farming Systems, the largest sustainable agriculture program in the United States, and a member of the Governor's Energy Policy Commission. Mr. Rich is on the board of Triangle Capital Corporation, a business development company operating throughout the southeast; Verenium, a cellulosic ethanol developer and enzyme production company; and he is a member of Cherokee Investment Fund's Sustainability Advisory Board. Mr. Rich is a 1967 graduate of Duke University.

JAMES A. ROTENBERG is Assistant Professor of Tropical Ecology at the University of North Carolina, Wilmington. His course on Global Environmental Studies focuses on several major current issues, including global warming. In 1998, he was awarded the J. William Fulbright Scholar to Guatemala. His primary research interest is tropical avian ecology, using birds as environmental research projects in Mexico, Belize, Costa Rica, Guatemala, California, and North Carolina. He received his Ph.D. in Biology from the University of California, Riverside.

WILLIAM SCHLESINGER is President of the Cary Institute of Ecosystem Studies. Before assuming this position in June 2007, he was James B. Duke Professor of Biogeochemistry, and Dean of the Nicholas School of the Environment and Earth Sciences at Duke University. He is the author or co-author of over 160 scientific papers and the widely-adopted textbook, *Biogeochemistry: An Analysis of Global Change,* Academic Press, 1997. He was elected a member of the American Academy of Arts and Sciences in 1995 and The National Academy of Sciences in 2003. He was President of the Ecological Society of America for 2003-04. Currently, his research focuses on global change ecology. Dr. Schlesinger has testified before U.S. House and Senate Committees on a variety of environmental issues, including preservation of desert habitats and global climate change. His Ph.D. is from Cornell.

CHRISTOPHER H. SCHROEDER is Charles S. Murphy Professor of Law and Professor of Public Policy Studies, and Director of the Program in Public Law at Duke University. His areas of research and scholarship include environmental and administrative law, democratic theory, legislative institutions, and separation of powers. He has written and taught about the regulation of toxic substances, the performance of American environmental policy and the philosophy of environmental protection, and is the author with Percival, Miller, and Leape of *Environmental Regulation: Law, Science and Policy*, now into its fifth edition (Aspen Publishers, 2006). He has served as Acting Assistant Attorney General in the Office of Legal Counsel at the Department of Justice and holds a J.D. from University of California, Berkeley.

DAVID A. SHEETS is a Senior Fellow with the Army Environmental Policy Institute, assisting the Army Secretariat in developing proactive policies and strategies to address emerging environmental, safety, and occupational health issues (ESOH) that may have significant impact on Army operational readiness. Mr. Sheets is responsible for sustainable energy initiatives and identifying future contaminants and munitions constituents of importance to the Army, to include nanoparticles and heavy metals. He is also responsible for several other initiatives, such as green chemistry and for improving the management of environmental cleanup. Mr. Sheets holds an MS in Environmental Health Engineering from the University of Texas at Austin and is a Licensed Professional Engineer in the State of Minnesota.

MARVIN S. SOROOS is Professor Emeritus of Political Science at North Carolina State University (Raleigh, North Carolina), where he has taught courses on global problems and policies and international law since 1970. Dr. Soroos has also been a pioneer in the field of global environmental politics, which has been his principal teaching and research interest since the mid-1970s. With David Orr, he co-edited an early book in the field entitled *The Global Predicament: Ecological Perspectives on World Order,* University of North Carolina Press, 1979. He is also the author of *Beyond Sovereignty: The Challenge of Global Policy,* University of South Carolina Press, 1986; and *The Endangered Atmosphere: Preserving a Global Commons,* University of South Carolina Press, 1997. His doctorate is from Northwestern University.

DENNIS TÄNZLER is a research fellow at Adelphi Research, an independent, non-profit institute that develops and implements innovative sustainable development strategies based in Berlin, Germany. His research focuses on climate and energy policies as well as on peace and conflict studies. In February 2007 he joined the Policy Planning Unit of the German Foreign Office as an expert on climate and energy policies. He has 10 years of experience in international environmental policy, European environmental policy and policy integration, and climate change and institutional aspects of environment, conflict, and cooperation. He has published widely on international affairs and environmental policy. He is the program director on the environment, conflict and cooperation platform (*www.ecc-platform.org*). He holds degrees in political science as well as in North American Studies, Cultural Sciences.

BERT B. TUSSING is Director, Homeland Defense and Security Issues, Center for Strategic Leadership, U.S. Army War College. He is a senior fellow with George Washington University's Homeland Security Policy Institute, the University of California, Irvine's Center for Unconventional Security Affairs, and Long Island University's Homeland Security Management Institute, and serves on the CSIS Beyond Goldwater-Nichols task force examining the government's response to catastrophic events. While in the U.S. Marine Corps, he participated in multiple humanitarian relief exercises, served as Marine Corps Analyst to the Secretary of the Navy and as Deputy Legislative Assistant to the Chairman of the Joint Chiefs of Staff. His publications include (with Antulio J. Echevarria) *From "Defending Forward" to a "Global Defense in Depth": Globalization and Homeland Security,* Strategic Studies Institute, 2003. Professor Tussing holds a Master's Degree in National Security and Strategic Studies from the U.S. Naval War College and in Strategic Studies from the U.S. Army War College.

SPENCER R. WEART is Director of the Center for History of Physics of the American Institute of Physics (AIP) in College Park, Maryland. Originally trained as a physicist, he is now a noted historian specializing in the history of modern physics and geophysics. He has written numerous articles and authored or co-edited seven books, most recently *The Discovery of Global Warming,* Harvard University Press, 2003 (revised edition in press). He has completed a variety of educational projects and has taught undergraduate and graduate courses on history of science at The Johns Hopkins University, the Eugene Lang College of the New School in New York City, and Princeton University.

He earned a Ph.D. in Physics and Astrophysics at the University of Colorado, Boulder, as well as a graduate degree from the University of California, Berkeley, in History.

ERIKA WEINTHAL is Associate Professor of Environmental Policy, Environmental Sciences and Policy Division at the Nicholas School of Environment, Duke University. She specializes in global environmental politics and natural resource policies with a particular emphasis on water and energy. The main focus of her research is the origins and effects of environmental institutions. She has written over a dozen articles in scholarly journals including, with others, "The Water Crisis in the Gaza Strip: Prospects for Remediation," *Ground Water*, Vol. 43, 2005, pp. 653-660; and is the author of the prize winning book, *State Making and Environmental Cooperation: Linking Domestic and International Politics in Central Asia*, MIT, 2002. She holds a Ph.D. in Political Science from Columbia University.

RICHARD WEITZ is a Senior Fellow and Director, Project Management at Hudson Institute. He is currently collaborating with the Center for Strategic and International Studies (CSIS) on a study of climate change and security. Dr. Weitz analyzes mid- and long-term national and international political-military issues, including employing scenario-based planning. He is the author of *Revitalising US–Russian Security Cooperation: Practical Measures,* International Institute for Strategic Studies, Adelphi Papers 377, 2005; and has published widely in journals and internet publications. He holds graduate degrees from the London School of Economics (M.Sc. in International Relations), Oxford University (M.Phil. in Politics), and Harvard University (Ph.D. in Political Science).